AN IDEA
WHOSE TI___

Since the French R̶e̶v̶o̶l̶___ ___ ___ ___ em has
played an increasingly dominant role in science and
international trade, and it was inevitable that we
would join the rest of the world in utilizing this
system of measures.

Mastering the Metric System clearly and thoroughly
explains metrics in all its many aspects. It gives the
historical background, lists the symbols, demon-
strates the mathematical logic, offers comprehensive
conversion tables, and shows how metrics works in
every area of measurement from weight to distance
to energy. Fully considered are the problems of
switching to the metric system and of educating both
students and the general public to use metrics properly. As a special added feature, this book includes
the bill recently passed by the Congress to begin
conversion to the metric system on a national basis.

The metric system has already changed our lives
somewhat, and will do so even more in the future.
Here is the book that contains all the information
you need for—

MASTERING THE METRIC SYSTEM

DR. NED V. SCHIMIZZI is Associate Professor of Education, State University of New York, College at Buffalo.
He has published numerous articles about teaching, and has
taught at both the elementary school and college levels.
Recently he has been active in designing experimental
curriculum, and has developed courses for teaching the
metric system to educators, professionals, and students. He
is listed in *Leaders in Education* and *Who's Who in the
East, 1975*.

Mastering the Metric System

by
Ned V. Schimizzi

A MENTOR BOOK
NEW AMERICAN LIBRARY
TIMES MIRROR
NEW YORK AND SCARBOROUGH, ONTARIO
THE NEW ENGLISH LIBRARY LIMITED, LONDON

MENTOR TRADEMARK REG. U.S. PAT. OFF. AND FOREIGN COUNTRIES
REGISTERED TRADEMARK—MARCA REGISTRADA
HECHO EN CHICAGO, U.S.A.

SIGNET, SIGNET CLASSICS, MENTOR, PLUME and
MERIDIAN BOOKS are published *in the United States* by
The New American Library, Inc., 1301 Avenue of the Americas,
New York, New York 10019,
in Canada by The New American Library of Canada Limited,
81 Mack Avenue, Scarborough, Ontario MIL IM8,
in the United Kingdom by The New English Library Limited,
Barnard's Inn, Holborn, London, E.C. 1, England

First Mentor Printing, August, 1975

5 6 7 8 9 10 11 12 13

PRINTED IN THE UNITED STATES OF AMERICA

DEDICATED to all those who wish to learn the metric system and to those enlightened minds of the past who recognized the need for it, who developed it, and who implemented it in the face of all forms of condemnation.

CONTENTS

CHAPTER I

THE HISTORICAL BACKGROUND OF THE METRIC SYSTEM

No cause, since the earliest organization of civilized society, has contributed more largely to embarrass business transactions among men, especially by interfering with the facility of commercial exchanges between different countries, or between different provinces, cities, or even individual citizens of the same country, than the endless diversity of instrumentalities employed for the purpose of determining the quantities of exchangeable commodities. For the inconvenience and confusion resulting from this cause, but one effectual remedy can possibly be suggested; and that is the general adoption throughout the world of one common system of weights and measures.

—*Frederick A. P. Barnard,*
President of Columbia College,
in an address to New York State educators,
August 1, 1871

European Adoption of the Metric System

Eighteenth-century Europe was a confusion of weights and measures. The units differed in name and value not only in neighboring countries, but also within the same country.

1

Commerce was handicapped, and both buyer and seller suffered.

The problems were not new, nor were possible solutions. Early decimal systems of measures were devised by the Babylonians, Phoenicians, and early Indians as well as early Europeans. However, as civilizations rose and fell, a confusing profusion of units developed. The advantages of a decimal division of measures were described in a brief pamphlet titled *La Disme* ("The Tenth Part") by one Simon Stevinus, an inspector of dikes in the Netherlands, in 1585. Numerous suggestions were made by leaders of many nations. In approximately 1660, the Royal Society of London proposed a project for decimal measures. In France ten years later, Gabriel Mouton proposed a comprehensive decimal system of measurement based on the "milliare"—the length of one minute of arc of a great circle of the earth.

Although nothing came directly from these proposals, two of the basic principles of the future metric system were being developed in them: a decimal division of measures and a natural scale of length. The two major themes were destined to recur for the next 120 years.

All efforts toward the adoption of a simplified international system of weights and measures failed until the French Revolution in 1789. It has often been suggested by historians and writers that such a pervasive element of daily life as a system of weights and measures could be successfully reformed only by so great an upheaval of events as a revolution in a major world power. Certainly the dawn of an era of reform and revision—of scientific methods as well as social systems—occurred in France.

At the time of its revolution, France had as bizarre a system of weights and measures as any nation in Europe. Not only were unit names superfluous and confusing, but standards differed in size from one locale to the next. At last, in May of 1790, a bill was passed which favored the development of a brand-new system of weights and measures. With this cue the French Academy of Sciences reported on its efforts to secure an internationally acceptable unit of length. A decimal fraction of the earth's circumference was recommended: specifically, the ten-millionth part of the meridianal distance from the north pole to the equator, through Paris. This unit became known as the "meter," which is rooted in the

Greek word *metron*, "a measure." Because of its similarity to already existing common European measures of that time, such as the ell, the yard, and the braccio, the meter was considered appropriate. The recommendation also proposed that the meter should become the only unit of length for all the world. All lengths from the shortest to the longest were to be expressed in meters and decimals of meters. A list of names for decimal multiples and submultiples was suggested.

At the same time, a new single unit of weight with decimal multiples and submultiples was proposed to replace the existing nondecimal system of standard weights. The new standard was derived from the weight of a cubic meter of water (or of some decimal part of the same). This began the formation of an entirely new system of weights and measures depending on the meter, later to be known to the world as the metric system.

In 1791 the French National Assembly voted to accept these recommendations. Since the distance from pole to equator was to serve as the foundation for the new system, this quarter-meridian had to be accurately measured, and the French Academy of Sciences was assigned the task.

It was a difficult one. The surveyors had to work with the greatest available precision to measure the rough mountainous and inaccessible areas between Dunkirk and Paris by triangulation, and to compute from this the length of the entire quadrantal arc from the North Pole to the equator, allowing for the deviation of the earth's actual form from a true sphere. The ten-millionth part of the total computed length had then to be accurately marked off on a suitable number of meter bars. These bars had to be copied for wide distribution. This incredible task of establishing the standards as well as constructing copies of precise models of each measure for distribution had to be undertaken and completed at a time when the nation was in the throes of a political revolution.

The committee of twelve academicians, which included the most reputable mathematicians, geodesists, and physicists of that day, met with King Louis XVI at the Royal Palace on June 19, 1791, to discuss this monumental undertaking. The king also asked Jacques Dominique de Cassini IV to attend. Cassini IV was the fourth in his line to serve as director of the Paris Observatory; his great-grandfather had been summoned from Italy by Colbert, minister to Louis XIV, in 1621. When he was asked why he wanted to measure the Paris meridian over again even though his father and grandfather had already done so in previous years, Cassini IV replied that he hoped to improve upon the recordings of his forefathers by using improved modern instruments for triangulation.

It is often noted that this interview of the king with the committee of the academy to arrange plans for developing the basis for a metric system may have been his last official act. He secretly attempted to cross the frontier and leave France the next day with his queen, Marie Antoinette. The royal fugitives were, of course, recognized and arrested at Varennes, returned to Paris, imprisoned, and executed.

The work of triangulation between Dunkirk and Barcelona proved to be dangerous and hazardous, carried on, as it was, within this reign of terror. Much of the work completed on the "firing line" was done by two surveyor-engineers, Delambre and Mechain. These two men were often arrested as spies and otherwise harassed throughout the project. It took them six years (1792–1798) to survey from Barcelona to Dunkirk. With this work as a basis, they were able to calculate the remainder of the quarter-meridian from the North Pole to the equator. They achieved incredible results considering the difficulties they had to surmount.

Before the work was completed, the metric system was adopted, in principle, by the new French Republic in April 1795. This was the first official adoption of the metric system by any country. A provisional standard meter, from the geodetical data available to date, was defined. The prefixes for decimal multiples and submultiples were defined: 10, 100, 1,-000, and 10,000 were assigned the Greek prefixes deca-, hecto-, kilo-, and myria-; 1/10, 1/100, and 1/1000 were assigned the Latin prefixes deci-, centi-, and milli-. "Liter" (French) was the name assigned to the volume of a cubic decimeter.

In 1799, a standard meter and kilogram were constructed, based on the work of the surveyor-engineers. The previous year, the French government had invited friendly European powers to send scientific delegates to Paris for participating in the work of defining proper values to the standard meter and standard kilogram. Eleven delegates responded, representing Spain, Denmark, and seven then European republics. The ten French scientists who were their hosts presented the work already completed by the commission, which included the metric unit of mass, called the gram and defined as the mass of 1 cubic centimeter (a cube that is 1/100 of a meter on each side) of water at its temperature of maximum density. The cubic decimeter (a cube 1/10 of a meter on each side) was chosen as the unit of fluid capacity, and this measure was given the name "liter." The entire group studied the problem and reported in 1799 when the standard meter and standard kilogram were defined. The legislative assemblies accepted the report of the commission, and the standard platinum meter bar and kilogram were deposited in the archives of France; they were later moved to the Paris Observatory.

The metric system did not receive total acceptance in France until much later. A major obstacle was Napoleon's imperial decree in 1812 that the metric system and the old French system exist side by side. The metric system was to be used in schools and in government offices, and the old French system was to be used in retail trade. The results of this decree was chaos. The decree of 1812 was repealed in 1837 by both houses of the French Parliament. The use of the decimal metric system was made law, with a system of fines for violators, and the general adoption of the metric system in retail commerce in France began.

Just as the Romans succeeded in spreading Greek culture throughout the civilized world, the Emperor Napoleon is often given the credit for spreading the metric system in Europe through his military conquests. Among Napoleon's activities in conquered nations was an attempt to Gallicize them, economically and culturally. In the early part of the nineteenth century, this meant, among other things, the introduction and use of the metric system.

Those European countries adjacent to France began to adopt the metric system in the first quarter of the nineteenth century. Belgium, Holland, and Luxembourg, together, adopted metrication in 1820. Switzerland began metricating in 1822 but was still doing so in 1877. Spain began in 1860, Italy in 1861, Germany and Portugal in 1872, Austria in 1876, Norway in 1882, Yugoslavia in 1883, Rumania in 1884, Sweden in 1889, Bulgaria in 1892, Denmark in 1912, Greece in 1922, and the Soviet Union in 1922. By 1925, the number of persons in Europe and Asia that lived in nations that had officially adopted the metric system exceeded 300-000 000. A total of thirty-five nations, including the major nations of Europe, Asia, and South America, had officially adopted the metric system.

British adoption of the metric system was still some years in the future, but of course it eventually happened—perhaps because metrication was a prerequisite for membership in the European Common Market. There is an irony here; for centuries the British had been secure in the knowledge that foreigners were at an automatic disadvantage in any business transactions with them. They were sheltered behind a system of weights, measures, and coinage that assigned 240 pence to the pound, 112 pounds to the hundredweight, 22 yards to the chain, and nine gallons to the firkin, and they had become accustomed to beating all the world at trade.

Today, with the exception of the United States and a handful of small countries, the entire planet is using or planning to use the metric system, and the United States has taken definite official steps toward adopting it.

Continuous Refinement of the Metric System

By the late 1860s, improved metric standards were needed to keep pace with scientific advances. In 1875, an international treaty, the "Treaty of the Meter," established well-defined metric standards for length and mass. A system was developed to facilitate recommendations and further refinements in the metric system. This treaty was known as the Metric Convention and was signed by seventeen countries, including the United States.

The permanent supreme authority established by the treaty to govern the official international use of the metric system is known as the General Conference of Weights and Measures. This group met for the first time in 1889 and at six-year intervals thereafter. The active organ of the conference is the International Committee of Weights and Measures (CIPM), which is composed of eighteen members and meets a minimum of every two years. The International Bureau of Weights and

Measures was then given facilities near Paris where complete modern laboratories are provided physicists and metrologists for the continuing improvement of the precision in defining metric units and in reproducing its standards. Essentially, this bureau monitors the standards of all metric nations to maintain fundamental units within established tolerances.

The Metrication of Science

An outline of the metrication of science might begin with the German physicists Gauss and Weber, who chose the millimeter, the milligram, and the second as the fundamental units used for their electro-fields studies. Later the steam engine and the rise of electrical and mechanical engineering led to the ohm, the volt, the ampere, and the farad. In 1901, the Italian physicist Giorgi developed a system for units of electricity and magnetism. This system was later adopted to fit the present international system. More recent entries to the metric sciences include photometry, the measurements of the intensity of ionizing radiations, and the atomic definitions of the second.

The Early Metric Movement in America

The colonies may have had some standards for basic measurements, but without uniformity. The standards had been imported at different times from different places and had the advantage of "different" accuracy; most were as reliable as a free paint-store yardstick. All these were then tempered with a healthy variety of colonial regulations, which were in turn accompanied by as many colorful interpretations.

The framers of the U.S. Constitution fulfilled a major responsibility to rationality by recognizing the need for a uniform national system of weights and measures. The creation of this system was made a specific responsibility of the newly formed Congress, which was given the authority "to coin money, regulate the value thereof, and of foreign coin, and fix the standards of weights and measures." The urgency for such a system was expressed by George Washington in his first inaugural address: "Uniformity in the currency, weights, and measures of the United States is an object of the greatest importance, and will, I am persuaded, be duly attended to."

Thomas Jefferson's plan for decimal coinage was adopted by Congress because "The most easy ratio of multiplication and division is that by ten. Everyone knows the facility of decimal arithmetic." According to the *Gazette of the United States* in 1789, "The Congress division of monies being an exact decimal method, admits of multiplication and division by only placing dots or taking them away. It is the quickest, most certain and easy method of reduction, both for the learned and the unlearned."

The new nation also sought consultation from Thomas Jefferson concerning matters of weights and measures. Jefferson, capable as he was, submitted two plans in an attempt to "define and render uniform and stable" the existing English system by firmly establishing units of length, area, volume, weight, and force—e.g., a single gallon of 270 cubic inches, a bushel of 8 gallons or 2 160 cubic inches, and so on.

Jefferson's second plan was far more rational. He proposed a whole new system of weights and measures based on decimal ratios, the same as the nation had recently adopted for its coins. His new "foot" was to be based on the "second's pendulum" which was to be multiplied and divided into units of ten, all units of weights and measures being derived from this unit. To avoid confusion, Jefferson would retain some of the old names for frequently used units. Jefferson was aware of the development of the metric system in France. His system coincided with the French system in the direct relations of linear, weight, and units of volume with complete reliance on simple decimal arithmetic. Although Jefferson's report was accepted by Congress, neither plan was adopted.

While Secretary of State, John Quincy Adams was asked to prepare a "Report upon Weights and Measures." The result, eloquently written and book length, took four years to complete and reviewed the progress of weights and measurements from biblical eras onward. Adams called attention to five features of the metric system which he considered to be advantages:

1. The "invariable" standard of length taken from nature.
2. The single unit for weight and the single unit for volume.
3. The decimal basis.
4. The relation of weight units to French coinage.
5. The uniform and precise terminology.
   Specifically, he wrote:

> The single standard, proportional to the circumference of the earth; the singleness of the units for all the various modes of mensuration; the universal application to them of decimal arithmetic; the unbroken chain of connection between all weights, measures, moneys and coins; and the precise, significant, short, and complete vocabulary of their denominations; altogether forming a system adopted equally to the rise of all mankind; afford such a combination of the principal of uniformity for all the most important operations of the intercourse of human society; the establishment of such a system so obviously tends to that great result, the improvement of the physical, moral, and intellectual condition of man upon earth; that there can be neither doubt nor hesitancy in the opinion that the ultimate adoption and universal, though modified, application of that system is a consummation devoutly to be wished.

Although Adams personally desired the establishment of a uniform international system of weights and measures, he was aware that at that time the constitutional rights of the states were just beginning to be examined by the Supreme Court and that any attempt to impose the metric system might be politically disturbing:

> The power of the legislator is limited over the will and actions of his subjects. His conflict with them is desperate, when he counteracts their settled habits, their established usages, their domestic and individual economy, their ignorance, their prejudices, and their wants: all which is unavoidable in the attempt to change, or to originate, a totally new system of weights and measures.

Another reason why Adams and Congress hesitated to adopt the metric system was that "it would be hazardous to deviate from the practice of Great Britain." One hundred years later the excuse given by the British for not adopting the metric system was that they did not wish to deviate from the practice of the United States, their biggest trading partner.

Irony of ironies—although the proposals of Jefferson and Adams were turned down, one Ferdinand Hassler, an employee of the Treasury Department, had his accepted. Believing that the Treasury Department had constitutional authority to establish standards for the customs, he proceeded in 1832 to define a complete set of standards for weights and measures for the United States. A set of standards was sent to each customs house. Congress in effect endorsed his work in 1838 by ordering the Secretary of the Treasury to supply a "complete set of all weights and measures adopted as standards" to the governor of each state, "to the end that a uniform standard of weights and measures may be established throughout the United States." All states were eager to grant the new standards official status. Thus the United States, by 1850, now had a "system" of weights and measures, if chaos can be systematized.

After President Lincoln had formed the National Academy of Sciences to advise the government on all technical matters, the subject of metrication was again discussed in 1863. A committee led by a physicist, Joseph Henry, was appointed at the request of the Secretary of the Treasury to study the status of standards for weights, measures, and coinage. After two years of study, the committee recommended the adoption of the metric system. The recommendation was approved by Congressman John A. Kasson (Iowa), chairman of the newly appointed House Committee on Coinage, Weights, and Measures. The Kasson committee's three metric bills were passed by Congress. The most important of these legalized the use of metric weights and measures. However, it also specified English-system equivalents of metric weights and measures. The Postmaster General was directed to place metric postal scales in post offices that exchanged mail with foreign countries. The Secretary of the Treasury was directed to furnish each state with one set of metric standards.

Congressman Kasson's intentions were clear. The metric system was not being made compulsory. Congress was merely permitting the use of the metric system while careful to maintain interest in reform. This tone concerning metrication was to become policy for many years.

A few years later, the University Convocation of New York asked Professor Charles Davies of Columbia College to head a committee for the purpose of investigating methods for promoting the advancement of the metric system. In 1871 he submitted a report recommending that nothing be done. He not only raised objections to the system but predicted dark consequences for the nation if it were adopted. However, Frederick A. P. Barnard, president of Columbia, rebutted and refuted Davies objections. He outlined a plan for educating citizens in the use of the metric system. Barnard proposed that it be taught in the schools and used in legislating the tariffs, for assessing customs duties, for public surveys, in military and naval establishments, and in post offices. In December 1873 an organization was formed, with Barnard as president, for the advancement of the metric system. In addition to advancing the metric system, the new American Metrological Society was concerned with internationally uniform coinage and standardized time zones. Among the influential members were Congressman Kasson, twelve or more U.S. Representatives and Senators, and many reputable scientists and educators from the colleges and universities.

This group later gave birth to the American Metric Bureau, founded in 1876 in Boston. Bernard was leader of this group as well. A young librarian, Melvil Dewey, was the group's executive director. He later became famous for his contribution to library science, the decimal system of classifying library books.

Meanwhile, the opposition refused to remain dormant and organized the International Institute for Preserving and Perfecting Weights and Measures. Founded in Boston in 1879 by Charles Latimer, an engineer, its major purpose was to preserve and perfect those weights and measures that were strictly Anglo-Saxon. The institute's philosophy was permeated by the movement known as "pyramidology"—the basic tenet of which was that the Great Pyramid at Gizeh, Egypt, was constructed by the hand of God and contained proof that the Anglo-Saxon race was one of the ten lost tribes of Israel. Thus, Anglo-Saxon weights and measures were of divine origin. The institute was not only opposed to other systems of measurements, it set forth to "purify" the

English system and eliminate all non-Anglo-Saxon influences. It seems that no one was aware that much of the colorful, picturesque English system of measuring was originally borrowed from the Romans. The word "inch" was derived from a Latin word meaning one-twelfth. It was defined in terms of thumb knuckles and later changed to the barleycorn definition.

The Treaty of the Meter

The Treaty of the Meter was signed by seventeen nations, including the United States, after a five-year meeting (1870–1875) in Paris. The Treaty of the Meter accomplished four important objectives:

1. It reformulated the metric system and refined the accuracy of its standards.
2. It provided for the construction of new measurement standards and distribution of copies to participating countries.
3. It established permanent machinery for further international action on weights and measures.
4. It established a world repository and laboratory, the International Bureau of Weights and Measures.

In 1889 the United States received its copies of the new measurement standards, including meter bars and kilogram weights. In 1893, by administrative order, the Secretary of the Treasury declared the new metric standards to be the nation's "fundamental standards" of length and mass, making the United States an officially metric nation. The yard, pound, and other customary units were to be defined as fractions of the standard metric units.

By signing the Treaty of the Meter, the United States joined with every other major nation of the world in endorsing the metric system as the international system of weights and measures. It also endorsed the machinery through which metric measurements are maintained at maximum accuracy. And still no major efforts were made to convert our nation to the system it had officially approved!

Another attempt to convert the nation was made in 1896 by U.S. Representative Dennis Henley of Brooklyn, who introduced a bill requiring all government departments to use the metric system and making it the only legal system recognized in the States by 1899. The bill was supported by the Committee on Coinage, Weights, and Measures and passed in the House by a vote of 119 to 117. But its opponents were organized, and the bill was returned to the committee, where it remained. Reports of the times claimed that most Congressmen feared adverse reaction from farmers and tradesmen in an election year.

A dozen other bills were proposed by 1910. Although support for the metric system continued to come from scientists, educators, and those directly concerned with standards, the opposition became even better organized. Two prominent spokesmen were Fred Halsey, a New York engineer, and Sam Dale, editor of a textile journal. Claiming the support of engineers, manufacturers, and workmen, they termed themselves "practical men" as opposed to "closet philosophers." Many of their arguments are still heard today by opponents of the metric system:

1. Engineering standards for nuts, bolts, and machine tools would have to be replaced at great cost and inconvenience.
2. The alleged simplicity of the metric system was illusory, because errors would be made through misplacing the decimal point.
3. Most of the world's commerce was being carried on in terms of English and U.S. units.
4. The government had no right to tell a man what weights and measures to use; and such laws would be unenforceable.

The Halsey and Dale opposition successfully blocked the implementation of the metric system until the Depression years with such propaganda publications as "What Real He-Men Think of the Compulsory Metric System," "Metric Chaos in Daily Life," and "Metric Nightmare." During this same period, the American Metric Association was formed. This group has survived to this day while enjoying the support of the academic and scientific communities as well as some manufacturers.

A thirty-year lull followed. After World War II the United States emerged so dominant in the manufacture and distribution of goods that there seemed no need for change. It was to be the last lull. Then the "Halley's Comet" of our time appeared to shake every fiber of our educational and scientific fabric: Sputnik. The mobilization that followed in education, science, and technology forced another serious consideration of the metric system because it was the language of scientific measurement. In 1957 the U.S. Army adopted the metric system for weapons and equipment. The Organization of American States proposed that the Metric System be adopted by the entire Western Hemisphere. In 1958 the United States agreed to define inch and pound units in metric equivalents. (It is a fact that the inch and the pound are defined by the meter and the kilogram.) In 1959 the acting Secretary of Commerce announced efforts to study the problems of metric conversion, and in 1960 the United States participated in the General Conference of Weights and Measures, which met for the purpose of further refining the metric system to better meet the needs of science.

Then, in 1965, the president of the British Board of Trade announced in Parliament that the United Kingdom would adopt the metric system over a ten-year period. This act left the United States as the only major nation still using an obsolete and chaotic system of weights and measures. Even the "inventors" of our system were discarding it.

In August 1968, our Congress passed the Miller-Pell Metric Study Bill, which became Public Law 90-472, authorizing the Secretary of Commerce to conduct the necessary investigation, research, and surveys to gauge the impact on this nation of the worldwide use of metric units and to determine our nation's course of action. The Metric Study Group of the National Bureau of Standards is to be commended for completing a comprehensive study which included all dimensions of our economy.

On September 14, 1970, NASA announced that "measurement values" used in the agency's future scientific-technical publications would be expressed in the International System of Units (SI),* known as the metric system. In 1971, Secretary of Commerce Maurice H. Stans presented the results of the study to Congress in the now famous report titled "A Metric America—A Decision Whose Time Has Come." The report was overwhelmingly in favor of the metric system.

*The abbreviation is of the French term *Système Internationale d'Unités*.

Finally, in 1973, a metrication bill was introduced in the U.S. Senate, and as of this writing, another bill—which, if it is passed, will be known as the Metric Conversion Act of 1975—is in committee. The text of the bill and comments on it will be found in the Appendix.

Thus except for the formalities, metrication is here—we have no alternative.

CHAPTER II

INTRODUCTION TO THE METRIC SYSTEM

What Is the Metric System?

The metric system is sanity. The metric system is teachable. The metric system is learnable. The metric system was intended by those who developed it to become an international system, "for all people, for all time." It is a permanent, accurate, universally understood system of standards. Historically, the foundation for the metric system was a unit of linear measures with which the weight of water could be correlated and decimalized. Because the unit of linear measure had to be a function of nature, it could be scientifically reproduced and verified, unlike some arbitrary standard that might be physically altered or destroyed.

All units in the metric system are related by the number 10. We might say that the metric system succeeds in simply and logically coordinating the measurements of length, area, volume, and mass via one decimalized system. The monetary currency of the United States, with its extraordinary convenience, was the first decimal system to be used nationally. The ratio between the units of the series (dollars, dimes, cents) is ten. Additions, subtractions, and other numerical operations are simple. Calculations with metric units require no conversion from unit to unit such as are necessary between inches and feet, ounces and pounds. In the metric system there is only one series of units for length, one for area, one for capacity, and one for mass.

In its original stages, the meter was the fundamental unit of the metric system. All units of length and capacity were to be derived directly from the meter, which was defined to be equal to one ten-millionth of the earth's quadrant. It was

17

originally intended that the unit of mass, the kilogram, should be identical with the mass of a cubic decimeter of water at its maximum density (4°C). However, the units of length and mass are now defined differently.

In October 1960, the Eleventh General International Conference on Weights and Measures proceeded to redefine the meter as equal to 1 650 763.73 wavelengths of the orange-red radiation in vacuum of krypton 86 corresponding to the unperturbed transition between the $2p_{10}$ and $5d_5$ levels.

The kilogram is now independently defined as the mass of a particular platinum-iridium standard, the International Prototype Kilogram, which is kept at the International Bureau of Weights and Measures in Sèvres, France.

Since 1964, the liter has been defined as being equal to a cubic decimeter. The meter is now a unit on which is based all metric standards and measurements of length, area, and volume.

The metric system is a soundly based, universally acceptable, universally understood, metrologically precise system of measurement which assigns each kind of property only one basic unit, thus allowing the user to choose from its decimal multiples and submultiples the particular unit which best suits his purpose by merely manipulating the decimal point. The metric system is not only sane, it is simple to use and simple to learn.

Why Use the Metric System?

Why use a system that three out of four Americans could not even begin to define in 1973? Perhaps the most effective way to begin answering this question is to define, in terms of their derivation, our customary English units of weights and measures.

Yard: the distance from the outstretched fingers to the tip of the nose of King Edgar, an obscure Anglo-Saxon king of a thousand years ago.

Foot: the distance covered by thirty-six barleycorns laid end to end.

Inch: the width of the thumb of some forgotten king, or three barleycorns laid end to end.

Mile: the distance a Roman soldier traveled in a thousand paces.

Fathom: the length of a Viking's outstretched arms.

Acre: the amount of land that could be plowed with a yoke of oxen in a day.

The derivations of the ton, pound, ounce, furlong, rod, gallon, quart, pint, and our other English units of measure are equally "colorful." This makeshift, sloppy system has driven generations of American schoolchildren (and their teachers) up the walls. It has tripled the tasks of engineers and draftsmen, confused shoppers, and encouraged fraud in places of commerce. According to consumer experts, processed foods are packaged in such a wide variety of sizes that it is next to impossible for a shopper in a supermarket to make a quick price comparison.

The metric system is not only based on precision, it epitomizes precision. At the same time, no other system of measurement can match its inherent simplicity. Although laymen need to know and use only a few simple concepts from it, it was deliberately designed to fill all the needs of scientists and engineers as well as laymen. Where other systems developed haphazardly, the metric system is logically streamlined, requiring the mastery of only six base units. Because metric is based on the decimal system, multiples and submultiples of any unit are always related by powers of ten and follow a consistent naming scheme. Calculations are so much easier and efficient that our aerospace industry alone would save approximately $65 million annually by using the metric system.

The metric system bridges the gap between measurement and computation. Distinct advantages (in or out of science) are (1) ease in conversion from one unit to another; (2) unification of measures of distance, mass and volume; (3) easy standardization of manufactured products of different countries.

The metric system has a more fundamental relationship to human anatomy because it is based on the number ten. We have ten fingers; and from the beginning of time people have learned to count on their fingers. Its efficiency is welcomed in a nation that makes 20 billion measurements each day.

We could improve our position as leader and merchant with the rest of the world by metrication. The restrictions of our customary system upon trade are costing us $10 billion to $25 billion a year. The savings in calculations alone will pay for the costs of conversion in a few years. U.S. citizens who travel and trade abroad would no longer be handicapped. The United States does $48 billion a year in export-import business with metric nations who increasingly demand metric measurements. Although small in relation to our total economy, our exports are a crucial factor toward the maintenance of a favorable trade balance in a metric world. Because our economy depends on trading raw materials, manufactured products, and technological ideas with metric countries, we are at a competitive disadvantage if we continue to use a measurement system different from that of the world market. If we want to continue to affect the setting of international standards of all sorts, especially in industry and technology, metrication is a requirement. Our military allies are metric; our conversion would simplify coordination and logistics. American companies that need to make metric products for sale in the United States or in foreign markets sometimes find it advantageous to build a plant in a metric country and employ persons already familiar with the metric system. The exporting of jobs to metric countries is already a problem.

The United States will benefit in all sectors of the economy by replacing its customary units with metric units. Our scientific research is conducted with metric units. Already our entire pharmaceutical industry has converted to metrics in both production and in marketing, with accompanying lower operating costs. Drug dispensing and patient statistics have been metricated in our hospitals. Elevations and distances in the firing of all weapons by the U.S. Army are now measured in meters. All basic triangulation by the U.S. Coast and Geodetic Survey is metric. The advantage of modular coordinated planning can be fully utilized in architecture and building construction by using the 100-millimeter basic module. Our present inch-pound engineering standards, such as the diameter of wire and the thickness of sheet metal, as well as the sizes of nuts and bolts, are incompatible with standards used elsewhere. This hinders or prevents the export of many U.S.

products. Even when a potential customer in another nation prefers a U.S. machine, he is less likely to import it if standard parts for repair and maintenance are not available in his country. Metric agreements among metric European countries in the electronics industry are keeping U.S. non-metric electronic products out. The USSR, China, and developing nations demand metric machinery, parts, and materials. We are at a disadvantage with metric Japan and Europe in attempting to benefit by supplying this demand.

Among the "measurement-sensitive" products in which dimensions are critical and in which we may be at a disadvantage when exporting by not metricating are computers, typewriters, vacuum pumps, clinical thermometers, and tractors. Metrication should encourage more efficient operations. When computing for metric conversion, engineers should see ways to weed out superfluous sizes of parts and materials. For example, during metric conversion in England, one manufacturing firm reduced its stock of fasteners (nuts, bolts, rivets) from 405 sizes to fewer than 200 sizes. Another firm replaced 280 sizes of ball bearings with only 30 metric types.

Because calculations can be made faster in metric, computers prefer to "talk" in metric—and we must talk to our computers. Computers have already drastically reduced the drudgery that is involved in translating one measurement language to another. Numerically controlled machine tools, increasingly used in manufacturing, are guided by computer programs. Converting to metric dimensions needs only a change in the program.

Finally, over 90 percent of the world's population now uses the metric system in everyday life, in manufacturing as well as in commerce and trade. We are a small island of antique measuring units in a metric sea. This sea is offering to us:

1. Ease of learning and understanding those increasingly complex interrelationships, as well as the interdependencies of our technological civilization.
2. Improved tools for thinking quantitatively and therefore more analytically.
3. A smooth and efficient manner for quantitative communication among people in weights and measures, and in concepts.

When Will We Begin Using the Metric System?

We have already outlined the course of the longest-running debate in the history of this country—whether or not the United States should convert to the metric system. But now the question of when or whether is unreal. The United States *is* going metric. The first deliberate steps in that direction have come in recent months—180 years after Thomas Jefferson, Secretary of State in 1790, suggested that the United States adopt the metric system of units that had been proposed for adoption by the French Republic. Had the Congress acted promptly and favorably on Jefferson's proposal at that time, the United States would have been the first nation to adopt the metric system instead of the last.

How Will We Begin Using the Metric System?

On the basis of the evidence gathered in the U.S. Metric Study, the Secretary of Commerce recommended that our nation change to the International Metric System through a coordinated national program over a period of ten years; at the end of this period the nation would be predominantly (but not necessarily completely) metric.

Each industry, the educational system, and other groups in our society will have the opportunity to work out their own timetables and programs tailored to meet their own needs. However, there will still be a federal coordinating body to assist those with problems related to timetables, programs, education, and technical matters. This federal coordinating body is to make decisions as to how the public can best be familiarized with the metric system. This agency may be similar to Britain's Metric Board, which was established with a built-in "cease to exist" date in order to assure that another government bureaucracy did not perpetuate itself. Britain's Metric Board provides education, information, materials, and coordination to the metric conversion program.

When we speak of metrication, we need to expect two kinds of changes: (1) a "soft" change—simply an exchange of one measurement language for another, such as the TV weather announcer reporting the temperature in degrees Celsius instead of degrees Fahrenheit; (2) "hardware" changes, such as the altering of sizes, weights, and other dimensions of physical objects. For example, the milk distributors must modify machinery to fill and seal liter containers instead of quart containers. "Soft" conversions do not present the problems accompanying "hardware" conversions. The output of some machines can now be converted to metric dimensions by adjusting a dial. However, a hydraulic pump required for the same machine may not be built to metric standards. In this case the pump would be maintained and repaired until it wore out, then it would be replaced by a new pump built to metric specifications. Persons making the change to metric units will make whatever "soft" and "hardware" changes as are necessary to do their work or to comprehend what is being said by the mass media, newspapers and television. For industrial engineers, factory workers, carpenters, surveyors, building inspectors, butchers, schoolteachers, and people in almost every walk of life, going metric will mean accepting the metric system as the primary and preferred system of measurement and ultimately thinking primarily in metric terms instead of in traditional English terms.

Some measurements and some dimensions will never change. We are not going to tear up all our railroad tracks for the sake of relating them to a rounded metric gauge. Some measuring units that are not a part of the International Metric System may be used wherever it is believed that communications and calculations are made clearer and easier by their use. Meteorologists may continue to use the term "bar"; astronomers may continue to use the term "light year." Existing buildings, aircraft carriers, railroad locomotives, power-generating plants, furnaces, and hair dryers will be replaced with new metric models only when they wear out or become obsolete. There will be no reason for requiring real-estate deeds in metric dimensions—i.e., meters instead of yards, and hectares instead of acres—until the property changes hands and/or is resurveyed.

Suggested solutions for the problems of metric changeover vary as much as our teaching methods. One very appealing proposal, which may also be the simplest method and the most likely to meet a friendly reception, is to use audio-visual techniques such as are now so successfully utilized on the TV program "Sesame Street." Creative educators everywhere are able and willing to implement such a program.

The need for the nation to convert to metric by planning and organizing rather than by no plan or by an abrupt and mandatory changeover is obvious. In a well-planned metric conversion program, some things will change rapidly, some slowly, and some never.

Who Will Use the Metric System Besides You?

The International Metric System will be used by manufacturing and nonmanufacturing disciplines, education at all levels, advertising, publishing, law, medicine, public health, agriculture, forestry, fisheries, agencies of federal, state, county, and local government, real estate, college athletics, finance, insurance, warehousing, transportation, construction, communications, retailers, wholesalers, chiefs of police, fraternal organizations, exporters, importers, home economists, professional and technical societies—in fact, all workers and all consumers; you and me. The ways conversion will affect some specific activities are discussed at length in Chapter VI.

CHAPTER III

METRIC PREFIXES, METRIC UNITS, POWERS OF TEN, AND SYMBOLS

Metric Prefixes

deci -= $\frac{1}{10}$ (or tenths)   deka- = 10 (or tens)

centi-= $\frac{1}{100}$ (or hundredths)   hecto-= 100 (or hundreds)

milli- = $\frac{1}{1\,000}$ (or thousandths) kilo- = 1 000 (or thousands)

There are three common metric prefixes for divisions of ten and three common metric prefixes for multiples of ten. These prefixes function with all metric units of measure.

Metric Lengths

centimeters	*instead of*	inches
meters	*instead of*	yards
kilometers	*instead of*	miles

Metric Weights

kilograms	*instead of*	pounds
grams	*instead of*	ounces

Metric Capacity

liters   *instead of*   quarts

Prefixes for *Multiples* of Ten

Deka- *is* 10:

1 dekameter	=	10 meters
1 dekagram	=	10 grams
1 dekaliter	=	10 liters

Hecto- *is* 100:

1 hectometer	=	100 meters
1 hectogram	=	100 grams
1 hectoliter	=	100 liters

Kilo- *is* 1000:

1 kilometer	=	1 000 meters
1 kilogram	=	1 000 grams
1 kiloliter	=	1 000 liters

Prefixes for *Divisions* of Ten

Deci- *means* 1/10:

1 decimeter	=	1/10 meter
1 decigram	=	1/10 gram
1 deciliter	=	1/10 liter

Centi- *means* 1/100

1 centimeter	=	1/100 meter
1 centigram	=	1/100 gram
1 centiliter	=	1/100 liter

Milli- *means* 1/1000

1 millimeter	=	1/1 000 meter
1 milligram	=	1/1 000 gram
1 milliliter	=	1/1 000 liter

Metric Equivalencies for Length

1 meter	=	10 decimeters
1 decimeter	=	10 centimeters
1 centimeter	=	10 millimeters
1 dekameter	=	10 meters
1 hectometer	=	10 dekameters
1 kilometer	=	10 hectometers
1 kilometer	=	1 000 meters

6 kilometers
    are 60 hectometers
        are 600 dekameters
            are 6 000 meters.

3 meters
    are 30 decimeters
        are 300 centimeters
            are 3 000 millimeters.

Metric Equivalencies for Weight

1 gram	=	10 decigrams	
6 grams	=	60 decigrams	
1 decigram	=	10 centigrams	
6 decigrams	=	60 centigrams	
6 grams	=	60 decigrams	= 600 centigrams
1 centigram	=	10 milligrams	

6 centigrams
    are 60 milligrams.

6 decigrams
    are 60 centigrams
        are 600 milligrams.

6 grams
    are 60 decigrams
        are 600 centigrams
            are 6 000 milligrams.

1 gram	=	10	decigrams
1 decigram	=	10	centigrams
1 centigram	=	10	milligrams
1 dekagram	=	10	grams
40 grams	=	4	dekagrams
20 dekagrams	=	200	grams
1 hectogram	=	10	dekagrams

80 dekagrams
    are 8 hectograms.

8 hectograms
    are 80 dekagrams
        are 800 grams.

1 kilogram	=	10	hectograms
30 kilograms	=	300	hectograms

3 kilograms
are 30 hectograms
are 300 dekagrams
are 3 000 grams.

1 kilogram	=	10 hectograms
1 hectogram	=	10 dekagrams
1 dekagram	=	10 grams
1 kilogram	=	1 000 grams

Metric Equivalencies for Capacity

Larger than a liter (multiples of)

dekaliter
hectoliter
kiloliter

Smaller than a liter (divisions of)

deciliter
centiliter
milliliter

Units larger than a liter

1 dekaliter	=	10 liters		
1 hectoliter	=	10 dekaliters	=	100 liters
1 kiloliter	=	10 hectoliters	=	1 000 liters

7 000 liters
are 700 dekaliters
are 70 hectoliters
are 7 kiloliters.

These are the same (equal):

        1 kiloliter and 10 hectoliters
        1 hectoliter and 10 dekaliters
        1 dekaliter and 10 liters
        1 kiloliter and 1 000 liters

Units smaller than a liter

1 liter is   =   to 10 deciliters
                        thus 30 deciliters are 3 liters.

1 deciliter is   =   to 10 centiliters
                        thus 5 deciliters are 50 centiliters

    2 liters
        are 20 deciliters
                are 200 centiliters.

1 centiliter is   =   to 10 milliliters
                        thus 6 centiliters are 60 milliliters.

6 liters
    are 60 deciliters
            are 600 centiliters
                    are 6 000 milliliters.

These are the same (equal):

        1 liter and 10 deciliters
        1 deciliter and 10 centiliters
        1 centiliter and 10 milliliters

Prefixes with Corresponding Powers of Ten

        10^{-3} milli-
        10^{-2} centi-
        10^{-1} deci-
        10^{0} unit (meter)
        10^{1} deka-
        10^{2} hecto-
        10^{3} kilo-

Powers of Ten with Place Values of Multiples of Ten and Divisions of Ten

10^3	=	1 000	thousands place
10^2	=	100	hundreds place
10^1	=	10	tens place
10^0	=	1	ones place
10^{-1}	=	1/10	tenths place
10^{-2}	=	1/100	hundredths place
10^{-3}	=	1/1 000	thousandths place

Metric Prefixes with Corresponding Powers of Ten and Decimal Place Value

Prefix	Power	Number-Decimal
kilo	10^3	1 000
hecto-	10^2	100
deka-	10^1	10
unit (meter, liter or gram)	10^0	1
deci-	10^{-1}	.1
centi-	10^{-2}	.01
milli-	10^{-3}	.001

Corresponding Place Values Between Our Number System and the Metric System

Number System	Metric System Prefix
thousands	kilo-
hundreds	hecto-
tens	deka-

ones	unit (meter, liter or gram)
tenths	*deci-*
hundredths	*centi-*
thousandths	*milli-*

Powers of Ten with Corresponding Metric Prefixes for Linear Measurement and Symbols

		Symbols
10^{-3}	meter is a *milli*meter	(mm)
10^{-2}	meter is a *centi*meter	(cm)
10^{-1}	meter is a *deci*meter	(dm)
10^{0}	meter is a meter	(m)
10^{1}	meters is a *deka*meter	(dam)*
10^{2}	meters is a *hecto*meter	(hm)
10^{3}	meters is a *kilo*meter	(km)

Multiples of Ten with the Meter

10 meters	=	1 dekameter
10 dekameters (= 100 meters)	=	1 hectometer
10 hectometers (= 1 000 meters)	=	1 kilometer

Divisions of Ten with the Meter

1/10 meter	=	1 decimeter
1/10 decimeter (= 1/100 meter)	=	1 centimeter
1/10 centimeter (= 1/1 000 meter)	=	1 millimeter

* Although dkm, dkl, and dkg are also acceptable symbols for deka-meter, dekaliter, and dekagram, the author chooses to use the symbols dam, dal, and dag to represent these units.

Computing Area in Meters

5 meters   ✕   5 meters   =   25 square meters
or 25 m^2

Metric Powers of Ten with the Liter (for Volume or Capacity) and Corresponding Prefixes and Symbols

10^{-3} liter is a *milli*liter   (ml)
10^{-2} liter is a *centi*liter   (cl)
10^{-1} liter is a *deci*liter   (dl)
10^{0} liter is a liter   (l)
10^{1} liters is a *deka*liter   (dal) *
10^{2} liters is a *hecto*liter   (hl)
10^{3} liters is a *kilo*liter   (kl)

Metric Powers of Ten with the Gram (for Weight or Mass) and Corresponding Prefixes and Symbols

10^{-3} gram is a *milli*gram   (mg)
10^{-2} gram is a *centi*gram   (cg)
10^{-1} gram is a *deci*gram   (dg)
10^{0} gram is a gram   (g)
10^{1} grams is a *deka*gram   (dag) *
10^{2} grams is a *hecto*gram   (hg)
10^{3} grams is a *kilo*gram   (kg)

Shifting the Decimal Point

Going from Kilometers to Meters

0.008 24   =   0.082 4   =   0.824   =   8.24
kilometer   hectometer   dekameter   meters

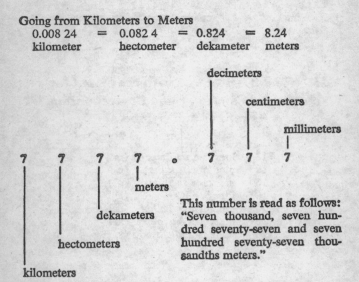

This number is read as follows: "Seven thousand, seven hundred seventy-seven and seven hundred seventy-seven thousandths meters."

A measurement of 4.62 meters would be read as "four and sixty-two hundredths meters"; that is, the 2 is centimeters, the 6 is decimeters, and the 4 is meters.

CHAPTER IV

THE INTERNATIONAL SYSTEM OF UNITS (SI)

The Seven Basic Units

The International System of Units (SI) is the refined modernized metric system established by international agreement by the Eleventh General Conference on Weights and Measures. The SI system is coherent in that the product or quotient of any two quantities in the system is the unit of the resultant quantity. This system provides a logical, unified system for all measurements in science, industry, and commerce. Its official abbreviation is SI. (As has already been explained, the abbreviation is derived from the French, Système International d'Unités.) The system rests on a skeleton of seven base units and two supplementary units. All other SI units are derived from these units. Multiples and submultiples are expressed by decimal manipulation. The table below presents "what" is to be measured along with the "unit" of measure to be used with each "quantity," and the "SI symbol" for the seven basic units and for the two supplementary units.

Quantity	Unit	SI Symbol
length	meter	m
mass	kilogram	kg
time	second	s
electric current	ampere	A
thermodynamic temperature	kelvin	K
amount of substance	mole	mol
luminous intensity	candela	cd

The Two Supplementary Units

Quantity	Unit	SI Symbol
plane angle	radian	rad
solid angle	steradian	sr

The Derived Units

Quantity	Unit	SI Symbol	Formula
acceleration	meter per second squared	...	m/s^2
activity (of a radio-active source)	disintegration per second	...	(disintegration)/s
angular acceleration	radian per second squared	...	rad/s^2
angular velocity	radian per second	...	rad/s
area	square meter	...	m^2
density	kilogram per cubic meter	...	kg/m^3
electric capacitance	farad	F	A.s/V
electrical conductance	siemens	S	A/V
electric field strength	volt per meter	...	V/m
electric inductance	henry	H	V.s/A
electric potential difference	volt	V	W/A
electric resistance	ohm	Ω	V/A
electromotive force	volt	V	W/A
energy	joule	J	N.m
entropy	joule per kelvin	...	J/K
force	newton	N	$kg.m/s^2$
frequency	hertz	Hz	(cycle)/s
illuminance	lux	lx	lm/m^2
luminance	candela per square meter	...	cd/m^2

Quantity	Unit	SI Symbol	Formula
luminous flux	lumen	lm	cd.sr
magnetic field strength	ampere per meter	...	A/m
magnetic flux	weber	Wb	V.s
magnetic flux density	tesla	T	Wb/m²
magnetomotive force	ampere	A	...
power	watt	W	J/s
pressure	pascal	Pa	N/m²
quantity of electricity	coulomb	C	A.s
quantity of heat	joule	J	N.m
radiant intensity	watt per steradian	...	W/sr
specific heat	joule per kilogram-kelvin	...	J/kg.K
stress	pascal	Pa	N/m²
thermal conductivity	watt per meter-kelvin	...	W/m.K
velocity	meter per second	...	m/s
viscosity, dynamic	pascal-second	...	Pa.s
viscosity, kinematic	square meter per second	...	m²/s
voltage	volt	V	W/A
volume	cubic meter	...	m³
wavenumber	reciprocal meter	...	(wave)/m
work (fxd)	joule	J	N/m

The Seven Basic Units Defined

The following definitions are included for the benefit of those who are interested in science and for students of elementary science. They are not important for laymen who are not interested in science and will not find it necessary to know specific scientific definitions.

1. Meter (m)—length

The meter is the standard unit of length in the metric system and is defined as 1 650 763.73 wavelengths in vacuum of the orange-red line of the spectrum of Krypton 86. The metric unit of area is the square meter (m²). Land is often measured by the hectare (10 000 square meters, or approximately 2.5 acres). The metric unit of volume is the cubic meter (m³). Volume of fluid is often measured by the liter (0.001 cubic meters).

2. Kilogram (k)—mass

The kilogram is the standard unit of mass. It is the only base unit which is still defined by artifact: the standard kilogram is based on a cylinder of platinum-iridium alloy kept by the International Bureau of Weights and Measures in Paris, France. A duplicate of this cylinder is now stored at the National Bureau of Standards in the United States.

The metric unit of force is the newton. It is defined as the amount of force that, when applied for 1 second, will give a 1-kilogram mass a speed of 1 meter per second.

The metric unit for pressure is the pascal.

The metric unit for work and energy of any kind is the joule, which is defined as 1 newton times 1 meter.

The metric unit for power is the watt, defined as 1 joule per second.

kilogram	kg	
newton	N	$1 \text{ N} = 1 \text{ kg.m/s}^2$
pascal	Pa	$1 \text{ Pa} = 1 \text{ N/m}^2$
joule	J	$1 \text{ J} = 1 \text{N.m}$
watt	W	$1 \text{ W} = 1 \text{J/s}$

3. Second (s)—time

The second is the basic unit of time in the metric system and is defined as the duration of 9 192 631 770 cycles of the radiation associated with a specified transition of the cesium 133 atom. It is realized by placing a cesium source at one end of an atomic beam spectrometer and bombarding the source until its atoms begin to travel at resonance frequency. One set of magnets in the spectrometer causes the atoms to travel in wave form while an oscillator tuned to the resonance frequency permits those atoms traveling at this frequency to pass through that detector which is at the end of the spectrometer. Another set of magnets deflects those atoms not traveling at proper speed. The detector then records the peak of each cycle as it passes. When 9 192 631 770 cycles are recorded, one second has passed.

The second was originally defined according to the earth's rotation. Early in the twentieth century it was discovered that the earth's rate of rotation varies, and a more dependable standard was needed. The present standard for the second was adopted in 1956.

The number of periods or cycles per second is called frequency. The SI unit for frequency is the hertz. One hertz is equal to 1 cycle per second.

The SI unit for speed is the meter per second.

NOTE: Standard frequencies and correct time are broadcast from WWV, WWVB, and WWVH, and stations of the U.S. Navy. Some shortwave radios receive WWV and WWVH on frequencies of 2.5, 5, 10, 15, and 20 *megahertz*.

4. Ampere (A)—electric current

The ampere is the basic unit of electric current in the metric system. It is determined by passing current through two parallel wires separated by a distance of 1 meter and then measuring the force of attraction between the wires, caused by their magnetic fields. The ampere is defined as that amount of current that will produce a force of exactly 2×10^{-7} newtons between the two wires for each meter of length.

ampere	A	
volt	V	$1 \text{ V} = 1 \text{ W/A}$
ohm	Ω	$1 \text{ }\Omega = 1 \text{ V/A}$

5. Kelvin (K)—temperature

The kelvin is the thermodynamic scale used to measure temperature in the metric system. It is defined as the fraction 1/273.16 of the thermodynamic temperature of the triple point of water. The temperature 0 K is called "absolute zero." The kelvin scale has its origin—i.e., its zero point— at absolute zero. This is the point at which all atomic vibration ceases. The triple point of water (the temperature at which water exists in all three states—vapor, liquid, and solid) is 273.16 kelvins.

On the commonly used Celsius temperature scale, water freezes at about 0° C and boils at about 100° C. The °C is defined as an interval of 1 K, and the Celsius temperature 0°C is defined as 273.16 K.

The Fahrenheit degree is an interval of 5/9° C or 5/9 K. The Fahrenheit scale uses °F as a temperature symbol. The Celsius scale uses °C as its symbol.

The standard temperature of the triple point is reached by filling an empty glass cylinder (of certain specifications) with pure water. The cell is cooled until a mantle of ice forms around the reentrant well and the temperature at the interface of solid, liquid, and vapor is 273.16 K. Thermometers are then calibrated by being placed inside the triple-point cell. Both the Celsius and Fahrenheit scales are derived from the kelvin scale.

6. Mole (mol)—amount of substance

The mole is the standard metric unit for the amount of a particular substance. It is defined as the amount of a substance in a system that contains as many elementary entities as there are atoms in 0.012 kilograms of carbon 12.

When the mole is used, the elementary entities must be specified. They may be atoms, molecules, ions, or electrons. They may be other particles or specified groups of such particles. It would not be practical to speak of a mole of apples or a mole of screws because the number of atoms in 0.012 kilograms of carbon 12 is extremely large. Moles are used in chemistry and physics in reference to atoms and molecules.

7. Candela (cd)—luminous intensity

The candela is the basic metric unit of luminous intensity. It is defined as the luminous intensity of 1/600 000 of a square meter of the cone of light emitted by a black body that has been heated to 2 042 kelvins, the freezing point of platinum. A black body is defined as any object that absorbs all the light which shines on it. Because no light is reflected from a black body, the radiant energy is converted to heat, raising the temperature of the object. If, however, the black body is heated, light will radiate from it.

The metric (SI) unit of light flux (amount of light) is the lumen (lm). A source which has an intensity of light of 1 candela in all directions radiates a light flux of 4 π lumens. About 1 700 lumens are radiated from a 100-watt light bulb.

The Two Supplementary Units Defined

1. Radian (rad)—plane angle

The radian is the plane angle with its vertex at the center of a circle that is subtended by an arc equal in length to the radius.

2. Steradian (sr)—solid angle

The steradian is the solid angle with its vertex at the center of a sphere that is subtended by an area of the spherical surface equal to that of a square with sides equal in length to the radius.

Metric Measurements of Some Familiar Things

Metric Weight

apple	200 g
loaf of bread	500 g
telephone	1 650 g or 1.65 kg
jockey	50 kg
football player	100 kg

Metric Area

book of pocket matches	19 cm²
monetary currency	98 cm²

school textbook	280 cm²
longplaying record	700 cm²
newspaper	2 600 cm²

Metric Volume

jigger	30 ml
coffee cup	250 ml
pickle jar	500 ml
milk bottle	1 000 ml or 1 liter
canning jar	1 000 ml
unit of gasoline	1 liter

Metric Length

football	27 cm
chair from floor to seat	45 cm
table from floor to surface	76 cm
height of a man	1.8 m

We shall use:

> meters as we would feet and yards
> grams as we would ounces and pounds
> liters as we would pints, quarts, and gallons.

Miscellaneous Common Quantities

comfortable room temperature	18° C to 21° C
body temperature	37° C
world maximum temperature	58° C
world minimum temperature	−90° C
1 kilo of rice	5 cupfuls
starvation level	4 000 J
absolute zero	0 K (−273° C)
height of average man	1.7 m
circumference of earth	40 000 km
acceleration due to gravity	9.81 m/s
speed of light	300 Mm/s
speed of sound (s.t.p)	333 m/s
atmospheric pressure	1 bar
moon mean distance	384 000 km
sun mean distance	150 Gm (10⁹ m)

Prefixes Indicating Divisions and Multiples of Basic Units

Length

Divisions (Latin)

deci-		1/10	(.1)
centi-		1/100	(.01)
milli-		1/1 000	(.001)
10	decimeters	= 1 meter	
100	centimeters	= 1 meter	
1000	millimeters	= 1 meter	

Multiples (Greek)

deka-		10 times
hecto-		100 times
kilo-		1 000 times
1 dekameter	= 10 meters	
1 hectometer	= 100 meters	
1 kilometer	= 1 000 meters	

Weight

10 milligrams	= 1 centigram
10 centigrams	= 1 decigram
10 decigrams	= 1 gram
10 grams	= 1 dekagram
10 dekagrams	= 1 hectogram
10 hectograms	= 1 kilogram

Volume

10 milliliters	= 1 centiliter
10 centiliters	= 1 deciliter
10 deciliters	= 1 liter
10 liters	= 1 dekaliter
10 dekaliters	= 1 hectoliter
10 hectoliters	= 1 kiloliter

Metric and Decimal Relationships: Place Value

Metric and decimal relationship in terms of place value is shown in the figure below. The meter coincides with the ones place, with its multiples and divisions related as follows:

Decimal place names:	thousands	hundreds	tens	ones	.	tenths	hundredths	thousandths
Exponential form:	10^3	10^2	10^1	10^0	.	10^{-1}	10^{-2}	10^{-3}
Units of linear measure:	kilometers	hectometers	dekameters	meters	.	decimeters	centimeters	millimeters
Standard decimal numerals:	1	0	0	0	.			
		1	0	0	.			
			1	0	.			
				1	.			
				0	.	1		
				0	.	0	1	
				0	.	0	0	1

Summary Scale of SI Prefixes of the Metric System

The meter—the unit on which the metric system is based —is in scale with the things human beings normally encounter, such as rooms, buildings, and short distances. The common multiples (deka-, hecto-, and kilo-) and the common submultiples (deci-, centi-, and milli-) are sufficient to measure the largest and the smallest things that most of us need to measure. But some scientists and technicians require additional prefixes because they deal with things inhumanly small, such as the sizes of atoms, or inhumanly large, such as the weights of stars. The scale below shows how these extra-small and extra-large prefixes extend from the everyday range of millimeter to kilometer.

Multiples and Submultiples		*Prefix*	*Pronun-ciation*	*SI Symbol*
1 000 000 000 000	$= 10^{12}$	tera-	(*ter*-a)	T
1 000 000 000	$= 10^{9}$	giga-	(*ji*-ga)	G
1 000 000	$= 10^{6}$	mega-	(*meg*-a)	M
1 000	$= 10^{3}$	kilo-	(*kil*-o)	k
100	$= 10^{2}$	hecto-	(*hek*-to-	h
10	$= 10^{1}$	deka-	(*dek*-a)	da
BASE UNIT 1	$= 10^{0}$			
0.1	$= 10^{-1}$	deci-	(*des*-i)	d
0.01	$= 10^{-2}$	centi-	(*sen*-ti)	c
0.001	$= 10^{-3}$	milli-	(*mil*-i)	m
0.000 001	$= 10^{-6}$	micro-	(*mi*-kro)	μ
0.000 000 001	$= 10^{-9}$	nano-	(*nan*-o)	n
0.000 000 000 001	$= 10^{-12}$	pico-	(*pe*-ko)	p
0.000 000 000 000 001	$= 10^{-15}$	femto-	(*fem*-to)	f
0.000 000 000 000 000 001	$= 10^{-18}$	atto-	(*at*-to)	a

Rules and Recommendations for Writing the Metric Units

If you are an American reader and have been reading this book straight through from cover to cover, you have probably noticed that one convention unfamiliar to you has been used throughout: instead of separating three-digit groups by commas, in this book they are separated by a space, so that "one million" appears as 1 000 000, not as 1,000,000. It would be surprising if this unfamiliar convention has caused you any real difficulty; you have probably adjusted to it with hardly a thought, as you would to a new pair of glasses or a different pair of shoes.

There are several such adjustments that people in both the United States and Canada will have to make as they adopt the metric system, and for some years—perhaps forever—they will have to be able to switch back and forth from one convention to another. This is hardly a new problem, and it is hardly a serious one. A mystery-story addict in Arizona has no difficulty in reading the works of one of England's mystery-story writers, even though "theater," "check," and "gasoline" appear as "theatre," "cheque," and "petrol."

There are many variations in the notation of numbers and quantities that will probably persist. In some countries, commas are used for decimal points; in some, decimal points are used as commas. In context, these variations cause no more difficulty than the mystery-story reader experiences, because the mind quickly adjusts to the convention being used. Nevertheless, it would be desirable to achieve international, or at least national, agreement on certain conventions.

Some countries have already established conventions, usually trying to eliminate any ambiguities that their own conventions might have to foreign eyes and at the same time avoid any changes that might discourage or confuse their nationals. Thus, the recommendation of both the U.S. Bureau of Standards and the Canadian Standards Association is to drop the use of the raised dot as a symbol of multiplication, because it is used by some countries as a decimal point; and to adopt the use of the space rather than comma for separating three-digit groups, because the comma is also used by other countries as a decimal point.

There are a few small differences between conventions in the United States and Canada. Perhaps the most significant is the spelling of the word "meter." Canada, like Great Britain, and like France where the word was coined, spells it "metre." The United States, which has for generations modified "-re" spellings to "-er" spellings, will probably go on spelling it "meter," and the recommendation of the National Bureau of Standards is that though the spelling "metre" should be encouraged "in communications whose principal audience is the technical community," the spelling "meter" is desirable for general audiences because in a small way it will help minimize resistance to adoption of the metric system. There is a similar difference in the spellings "litre" and "liter," but this is considered less important, since the liter is not a base unit.

Another spelling difference exists for the prefix "deca-" or "deka-." "Deka-" is favored in the United States, and is certainly truer to the Greek; "deca-" is the spelling in Great Britain and Canada. "Deca-" was the French coinage, and perhaps therefore has greater legitimacy—but then, the French had little choice, for the letter k is all but absent from their language.

Some further differences are noted in footnotes to the rules which follow. They are all, of course, trivial—and SI remains a truly international system in spite of them.

1. The decimal point is in the same position as a period or a raised dot.*   3.2 or 3·2

2. Groups of digits are separated by a space. Commas are not used. A four-digit sequence *may* remain unbroken.†

    52 000
    1.121 201
    2 750 g
    0. 113 5 mm

*The United States will almost certainly continue to use the decimal in the period position, and, like Canada, will discourage use of the raised dot as a sign of multiplication.
†The replacement of comma by space seems easy enough, but there is a complication. After the decimal point, should one begin the three-

3. Symbols for units are in Roman upright.

4 kg

4. Symbols for physical quantities are written in *italic*.*

$F = mV$
$V = 1$ bh

5. Unit symbols do not have plurals. If the name of the unit is written in full, then the normal plural form is used.

45 m
45 meters

6. Do not leave a space between a prefix and a symbol.

cm
mm

7. If two or more unit symbols are combined, a space is left between them.†

W m
km h

8. A space is left between the numeral and the symbol.

35 kg
20 m

9. No period or other punctuation is used unless it is for an end of a sentence.

22 cm

---

digit grouping from the right or from the left? So far, publications of the Canadian Standards Board group them from the left, which means that the spaces are in mirror order to those on the left side of the decimal. Thus the value of pi in metric notation is in Canada written 3.141 59 rather than 3.14 159. In the United States, the Bureau of Standards, though its own publications have not been consistent, also seems to prefer to group them starting from the left, and that is the convention used in this book. This system shows the digital group more clearly; the other probably appears less odd to U. S. eyes. But in any case, the omission of any digit grouping after the decimal, which is common now, may continue for many applications after the adoption of the metric system.

*This typographic convention is common now in many countries, and will probably be observed or not observed as formerly. In this book, all symbols are in roman type.

†The Canadian Standards Association recommends using the slash (/, also called virgule or solidus) for derived units that are formed by division—that is, in spoken language, by the connective "per." Thus "kilometers per hour" could be written km/h. This notation is used in the *National Bureau of Standards Technical News Bulletin*, 57 No. 6 (June 1973), as well.

10. A symbol for a unit is always associated with a numeral. If there is no numeral, the unit is written out.

2.4 kW
Electricity is measured in kilowatts.

11. A zero should precede numbers that are less than 1.

0.25 m

12. Names of units, whether or not they are named after a person, are not usually written with a capital letter.

watt
ampere
celsius

13. The symbol for a unit named after a person has a capital letter.

W
A
°C

The most commonly used rules may be:
1. In the plural, the symbols remain the same, e.g., 60 g; no "s" is added and a space is left between the numeral and the symbol.
2. No period is placed after a symbol unless it is at the end of a sentence.
3. Groups of three digits are separated by a space; commas are not used. A sequence of four digits may remain unbroken, e.g., 53 000; 3720 or 3 720.

Some Rules and Recommendations for Speaking About Metric Units

1. The prefix is accented when it is combined with a unit name. Examples:

kill-oh-meter	—kilometer
kill-oh-liter	—kiloliter
mill-i-meter	—millimeter
mike-ro-meter	—micrometer

2. Decare and hectare, the units for measuring land, are pronounced *deck*-air and *heck*-tair.
3. The use of plurals is permissible and grammatical as long

as they are not carried over to written symbols; e.g., fifteen meters, forty kilograms.

4. In many countries the term "kilo" is used for kilogram. This term should not be used for "kilometer."

5. Digits placed after a decimal point should be spoken separately. An example would be 3.86 kg, "three point eight six kilograms."

6. The squares and cubes of units may be referred to in the following two ways:

"square centimeters" or "centimeter squared"
"square meters" or "meter squared"
"cubic meters" or "meter cubed"

The singular is used in the second form. For example, 6 square centimeters is the same as 6 centimeter squared, but is *not* the same as 6 centimeters squared; 6 centimeters squared is 36 square centimeters. In mathematical notation, the difference is plain: 6 cm^2 means 6 square centimeters, but $(6 \text{ cm})^2$ means 36 square centimeters.

7. The abbreviation "c.c." should not be used for "cubic centimeter"; it is not an official SI symbol. It may retain some unofficial status, however, since it has been used for many years to denote engine displacement, syringe volume, etc.

Comments on Using Prefixes

1. A prefix is written next to the unit for the purpose of avoiding confusion where spacing is necessary. For example, while ms^{-1} means "per millisecond," m s^{-1} means "meters per second."

2. Prefixes should not be used in combination; we do not say 1 m m m to mean 1 millimillimeter, but we say $1 \mu\text{m}$ for a millionth of a meter (1 micrometer).

3. When the size of a derived unit which is a quotient is changed, the prefix should be applied to the numerator. For example, the density of water is not 1 g/cm^3, but 1 Mg/m^3.

4. A quantity should be expressed by using one unit only.

Although it is not incorrect to express a weight as 4 kg 320 g, it is preferable to express it as either 4 320 g or 4.32 kg.

5. When a prefix and a unit are raised to a power, the whole unit is raised and not just the basic unit; cm² is the same as (cm)²; not a hundred times a meter squared.

A Review of the Most Common Metric Measures for Everyday Use

Most people will need to learn only four metric measures. These are the:

meter	for length
liter	for liquids (or volume)
kilogram	for weighing
°C	for temperature

Remember that metric measures are a base-ten system and move up and down by tens:

kilo- means a thousand. A kilometer is a thousand meters.

centi- means a hundredth. A centimeter is a hundredth of a meter.

milli- means a thousandth. A millimeter is a thousandth of a meter.

Length in metric

The meter (m). 3 m means three meters.

The meter is the base metric measurement of length. It is slightly longer than 3 feet 3 inches. When measuring, measure directly in metric. Do not measure in feet and inches first and then attempt to convert. This is a waste of time.

The centimeter (cm). 2 cm means two centimeters.

The centimeter is used for measuring shorter lengths. It is one-hundredth of a meter. Two and one-half (2½) centimeters are nearly 1 inch.

The millimeter (mm). 7 mm means seven millimeters.

A millimeter is a very small unit. It is one-thousandth of a meter. It takes 25 millimeters to approximate 1 inch. The most-used metric measurement in building and engineering is the millimeter.

The kilometer (km). 10 km means ten kilometers.

The kilometer is the metric measure for distances. It is 1 000 meters or approximately 5/8 of a mile.

Liquids in metric

The liter (l). 4 l means four liters.

The liter is the basic metric unit for measuring liquids. It is slightly more than one quart. While 4.55 liters approximate one Imperial Gallon, there are 3.79 liters per U.S. gallon.

The milliliter (ml). 300 ml means three hundred milliliters.

One fluid ounce contains approximately 28 milliliters. The spoonful may be 5 ml.

The centiliter (cl). 50 cl means fifty centiliters.

A centiliter is a hundredth of a liter.

The deciliter (dl). 9 dl means nine deciliters.

There are 10 centiliters to the deciliter. One liter contains 10 deciliters.

Weight in metric

The kilogram (kg). 4 kg means four kilograms.

Metric weighing is in kilograms. A kilogram is slightly less than 2¼ pounds.

The gram (g). 250 g means two hundred and fifty grams— one quarter of a kilogram.

A thousand grams are the same as one kilogram. One ounce contains approximately 29 grams, and 454 grams approximate one pound.

The tonne (t). 10 t means ten tonnes.

A tonne is slightly less than a ton. It contains 1 000 kilograms. It is referred to as a metric tonne.

Temperature in metric

The everyday metric measurement of temperature is the degree Celsius. Celsius is the same as centigrade. It is written °C and spoken as "degree C."

The boiling point of water is 100° C and the freezing point is 0° C.

Those temperatures below freezing (0° C) have a minus sign. A cold winter night may be —5° C.

Normal body temperature is 37° C, corresponding to 98.4° F.

Centrally heated bedroom temperatures may be 18° C while living rooms may be 22° C.

A heatwave may be 30° C; 20° C is a warm day. A mild winter day or cool summer day may be 15° C. 5° C is cool. 0° C is freezing, and —5° C is a cold winter night.

CHAPTER V

LEARNING THE ELEMENTARY CONCEPTS OF LENGTH, AREA, VOLUME, AND CAPACITY IN THE METRIC SYSTEM

There are six elementary prefixes in the metric system.

Three of these prefixes are in Latin and are used for divisions of ten.

These three prefixes are:

1. _____ = $\frac{1}{10}$ (or tenths)

_____ = $\frac{1}{100}$ (or hundredths)

_____ = $\frac{1}{1\,000}$ (or thousandths)

Three of these prefixes are in Greek and are used for multiples of ten.

These three prefixes are:

2. _____ = 10 (or tens)

_____ = 100 (or hundreds)

_____ = 1 000 (or thousands)

These prefixes function with all metric units of measure.

Lengths

3. In the metric system, the meter is the unit of measure for length.

   If a metric stick is used to measure meters, then a meter stick is one _____ in length.

4. The meter shall replace the yard.

   _____ shall replace inches.

   _____ shall replace miles.

5. When a meter stick goes four times alongside a car, the car is four meters in length.

   If a meter stick goes five times alongside a car, the car is 5 _____ in length.

6. 2 m is the briefest possible method of symbolizing 2 meters.

   m is used because it is the initial letter of the word meter. Thus _____ is the briefest way to write 6 meters.

   Is there a period after 2 m?

7. Name three items for which the meter would be too long a unit for measuring. _____

   _____

   _____

8. The meter is divided into ten equal parts. Each part is called a _____ and is one _____ of a meter in length.

9. The briefest way possible to symbolize decimeter is dm.

> d is for deci-
>
> m is for meter

If 2 dm is the same as 2 decimeters, then _____ is the same as 3 decimeters.

10. The decimeter is divided into ten equal parts. Each part is called a _____ and is _____ of a decimeter in length; and _____ of a meter in length.

> "deci-" means "tenth."

11. If there are indeed ten centimeters in a decimeter and ten decimeters in a meter, then how many centimeters are there in a meter? _____

Thus, a centimeter is _____ of a meter.

> "centi-" means "hundredth."

Some things are too large to be measured by the decimeter.

12. The centimeter is symbolized by the letters cm.

> c is for centi-
>
> m is for meter

If 4 cm are 4 centimeters, then _____ are 7 centimeters.

13. If three meters (3 m) contain 300 centimeters (cm), then 6 m contain _____ cm.

14. If there are 200 cm in 2 m, then there are 400 cm in _____.

15. Some things may be too small to be measured by the centimeter. The centimeter, in turn, is divided into ten equal parts.

    We call each part a ――――――――――――

16. A centimeter contains ten millimeters. A meter contains one hundred centimeters.

    Thus there are ―――――――――――― millimeters in a meter.

    "milli-" means "thousandth."

    A millimeter is the same as ―――――――――― of a meter (give fraction and the decimal fraction).

17. The millimeter is symbolized by the letters mm.

    m is for milli

    the other   m is for meter

    If 5 mm are 5 millimeters, then ―――――――――――― are 8 millimeters.

18. One meter contains one thousand millimeters. This is symbolized numerically as follows:

    1 m contains 1 000 mm

    then   3 m contains ――――――――――――

19. If 1 m contains 1 000 mm and 3 m contains 3 000 mm then there are 4 000 mm in ――――――――――――

10	×	10	×	10
deci-		centi-		milli-
10ths		100ths		1 000ths

20. "Deka" is the Greek word for ten. When used as a prefix with meter, it means ten times a meter. Ten times a meter is a ――――――――――――

21. A dekameter is _____ meters.

"deka-" means 10 ×

22. There are 10 _____ in each dekameter.

Therefore, one meter is _____ of a dekameter.

23. The dekameter is symbolized by the letters dam.

da is for dekam-

m is for meter

If 4 dam are 4 dekameters, then _____ are 6 dekameters.

five dekameters are written as _____.

24. A hectometer is ten times a dekameter. It is used to measure objects which are too long to measure conveniently in dekameters. A hectometer contains ten dekameters. If "deka-" means "ten times," a hectometer means _____ times a dekameter.

25. The hectometer is symbolized by the letters hm.

h is for hecto-

m is for meter

Therefore, if 3 hm are 3 hectometers, then _____ are 8 hectometers.

26. If a dekameter contains ten meters, and a hectometer contains ten dekameters, then a hectometer contains _____ meters.

27. Hectometer means 100 × a _____.

28. One meter is _____ of a hectometer.

29. The kilometer is symbolized by the letters km.

    k is for kilo-

    m is for meter

    If 6 km are six kilometers, then _____ are three kilometers.

30. Kilometers are used for measuring long distances. Kilo means _____ × a _____.

31. There are _____ hectometers in a kilometer.

    A kilometer is _____ hectometers.

    A hectometer is _____ meters.

    A kilometer is _____ meters.

32. A kilometer is the same as _____ times a _____.

33. One meter is _____ of a kilometer.

34. 35 kilometers are written symbolically as _____.

35. 1 km contains 1 000 m. Therefore, 6 km contains _____

    5 423 m = 5 km   423 m.

    4 362 m = _____   _____

10	×	10	×	10
deka-		hecto-		kilo-
10		100		1 000

    The decimal is metric and mobile.

36. There are ten decimeters in a meter. Therefore, a decimeter is one tenth of a meter. However, in the metric system we use decimals to express fractions. We call these decimal fractions.

If 1 dm is expressed as 0.1 m, how do we express 5 dm as a decimal fraction? _____

37. If 5 m 3 dm is expressed as 5.3 m, then 4 m 2 dm is expressed as _____.

38. There are one hundred centimeters in a meter. A centimeter is one hundredth of a meter. One hundredth of a meter is expressed in a decimal fraction as 0.01 m.

If 1 cm is expressed as 0.01 m, then 4 cm is expressed as _____ and 6 cm is expressed as _____.

Have you ever noticed a mathematical notation in which the decimal was elevated? For example, 4·2 m. This is the convention in some countries. It should not be confused with the raised dot multiplication sign; it means "point," not "times."

Have you noticed that a zero precedes numbers that are less than 1?

Digits placed after a decimal point should be spoken separately. Example: 4.26 kg is "four point two six kilograms."

39. 7 m   3 cm is expressed in decimal fractions as 7.03 m.

Therefore 9 m   4 cm is expressed as _____.

Likewise, if 9 m   42 cm is expressed as 9.42 m in a decimal fraction, then 8 m   27 cm is expressed as _____.

Decimal fractions may be added, subtracted, multiplied, and divided.

40. If four meters by eight meters are thirty-two square meters, then 5 m × 6 m = _____.

A living room is 7 by 5 meters. In area it is _____.

41. Label each digit according to the metric unit it represents.

$$9 \quad 9 \quad 9 \quad 9 \quad \bullet \quad 9 \quad 9 \quad 9$$

Read the above numerical statement seven times.
Shift the decimal point before each reading.

42. The cubic meter is used to measure volume. A cubic meter is defined as that area of a cube, the sides of which are one meter in length.

The cubic meter is symbolized as m^3.

If 7 cubic meters are symbolized as 7 m^3, then 9 cubic meters are symbolized as ———————.

43. If the volume of a cube whose sides are one meter in length is 1 m^3, then a closet whose volume is eight times as much is symbolized by ———————.

The cubic meter is also prefixed with deci- and centi-. We have cubic decimeters and cubic centimeters.

44. The cubic decimeter is symbolized as 1 dm^3. It is defined as the area of a cube the sides of which are ————— in length (review no. 42).

45. If 4 cubic decimeters are symbolized as 4 dm^3, then 8 cubic decimeters are symbolized as ———————.

46. 2 dm^3 + 3 dm^3 + 4 dm^3 = ———————.

47. For smaller volumes the cubic centimeter is used. A cubic centimeter may be defined as the area of a cube whose sides are ————————— in length (review no. 42).

48. The cubic centimeter is symbolized by ———————.

49. If 2 cubic centimeters are symbolized as 2 cm^3, then 5 cubic centimeters are symbolized as ———————.

50. 7 cm³ + 2 cm³ — 4 cm³ = _____

51. To compute the volume of a cube (or to cube anything) one must multiply a given digit by itself 3 times.

    $4^3 = 4 \times 4 \times 4 = 64.$

    Thus, the volume of a cube with sides 4 cm in length is _____

52. If a cube has sides that are 3 cm in length, its volume is _____

Liquid Volumes

53. The liter is the basic metric unit for measuring liquids. It is the volume of a kilogram of distilled water at 4 °C. It is equal to 1 cubic decimeter, which is equal to _____ cubic centimeters.

    Therefore 1 liter is equal to _____ cm³.

54. The liter is symbolized as _____

    If 8 liters are symbolized as 8 1, 6 liters are symbolized as _____

55. If "deka-" is 10 and "hecto-" is 10 × 10, then "kilo-" is _____

56. If "deci-" is 1/10, then centi- is _____ and "milli-" is _____

57. If the liter were divided into ten parts, one of those parts would be a _____

    The symbol for deciliter is dl.

                         the d is for deci-

                         the l is for liter

58. Five deciliters would be symbolized as _____.

    5 dl + 3 dl − 1 dl = _____.

59. 1 dl expressed in a decimal fraction is _____ l.

60. 5 dl would be expressed in a decimal fraction as
    _____ l.

    0.2 l + _____ + 0.3 l = 0.9 l.

61. Five liters contain _____ deciliters.

62. If a deciliter is 1/10 of a liter, then a _____
    is 1/100 of a liter.

63. The centiliter is symbolized as cl.

             the c is for _____

             the l is for _____.

64. 4 centiliters are symbolized as _____
    and decimalized as _____ l.

65. 2 cl + 3 cl + 1 cl = _____
    (in decimal fraction).

66. If 365 cl = 3.65 l, then 402 cl = _____.

67. 3.65 l + 4.02 l are _____.

68. If a centiliter is 1/100 of a liter, a liter contains
    _____ centiliters.

69. If a liter contains 100 centiliters, 7 liters contain
    _____ centiliters.

70. 4 l + 12 cl = 412 cl.

    412 cl = 4 l + 12 cl.

    3 l + 11 cl = _____ cl.

71. If a deciliter is 1/10 of a liter and if a centiliter is 1/100 of a liter, then a _____ is _____ of a liter.

72. The milliliter is symbolized as _____.

The _____ is for _____,

the _____ is for _____.

73. Six milliliters are symbolized as _____

Eight milliliters are symbolized as _____.

74. If deci- is 0.1 and centi- is 0.01, then _____ is _____.

75. If 23 ml = 0.023 1, then 31 ml = _____.

76. If 4 385 ml = 4.385 1, then 2 416 ml = _____.

77. A liter contains: _____ deciliters,

_____ centiliters,

_____ milliliters.

78. A dekaliter contains _____ liters.

A hectoliter contains _____ liters.

A kiloliter contains _____ liters.

| 10 | × | 10 | × | 10 | = | 1 000 |
| deka- | | hecto- | | kilo- | | |

79. A dekaliter is _____ a liter.

A centiliter is _____ of a liter.

A milliliter is _____ of a liter.

80. If 5 1* 214 ml = 5 214 ml and 3 1 172 ml = 3 172 ml, then 6 1 112 ml = _____

* 1 is the symbol for liter.

Weights

Metric weighing is done in kilograms. A thousand grams are the same as one kilogram; kilo- means 10 × 10 × 10.

81. The kilogram is symbolized as _____.

    the _____ is for kilo-

    the g is for _____.

82. If 3 kilograms are symbolized by 3 kg, then 7 kilograms

    are symbolized by _____.

83. If one kilogram contains 1 000 grams, then 4 kilograms

    contain _____ grams.

84. If 2.3 kilograms are the same as 2 300 grams, then 4.6

    kilograms are the same as _____ grams.
    grams.

85. If 6.56 kilograms are 6 560 grams, then 9.25 kilograms

    are _____ grams.

86. "Hecto-" means _____.

87. A hectogram contains _____ grams.

88. The hectogram is symbolized as hg.

    the h is for _____

    the _____ is for grams

    hg is for _____

    hg contains _____ grams

    1 000 grams are one _____.

89. "Deka-" means _____, therefore a dekagram

    is _____ grams.

    Ten grams are one _____.

90. The dekagram is symbolized by dag.

        the da is for _____

        the g is for _____

91. dag is for _____.

    "deka-" means _____.

    In a dekagram there are _____ grams.

92. If 5 dekagrams are symbolized by 5 dag, then 7 deka-

    grams are symbolized by _____.

93. For very small or lightweight objects we use the gram.

    A gram is symbolized by g. Three grams are symbolized
    by 3 g.

    4 grams are symbolized by _____.

94. 1 gram is _____ of a kilogram.

95. In decimal fractions this would be expressed as 1 g =
    0.001 kg.

    Therefore 2 g = _____ kg

           25 g = _____ kg

        275 g = _____ kg

96. If 4 kg 2 g = 4.002 kg, then 3 kg 4 g = _____ kg.

97. The prefix for 1/10 is _____.

The symbol for gram is _____.

One tenth of a gram is a _____.

The decigram is symbolized by _____.

If 1 g = 10 dg, then 2 g = _____ dg.

98. Six decigrams are _____ centigrams

are _____ milligrams.

99. Six grams are _____ decigrams

are _____ centigrams

are _____ milligrams.

100. The centigram is symbolized as cg.

the c is for _____

the g is for _____

centi- means _____.

101. One hundredth of a gram is a _____.

102. If 3 centigrams are symbolized as 3 cg, then 5 centigrams can be symbolized as _____.

103. There are 100 cg in every _____.

Therefore 7 g = _____ cg

and 600 cg = _____ g.

104. And what prefix comes after centi-? milli-, of course.
For even lighter weights we use a milligram.

The prefix milli- means _____; therefore a milligram is _____ of a gram.

There are _____ milligrams in a gram.

105. The milligram is symbolized by mg.

the m is for _____

and the g is for _____

106. If 7 milligrams are symbolized by 7 mg, then 9 milligrams are symbolized by _____.

107. If 1 g = 1 000 mg, then 2 g = _____ mg.

   1 g + 4 g + 2 g = _____ mg.

   7 000 mg = _____ g.

108. In decimal fractions, 0.1 is one tenth,

   0.01 is one hundredth,

   0.001 is one thousandth.

   Therefore 1 mg = _____ g

   and 6 mg = _____ g.

109. In decimal fractions:

   1 g = _____ kg

   1 kg = _____ g

   45 g = _____ kg

   45 kg = _____ g.

110. 1 dg = _____ g.

   1 g = _____ dg.

111. 1 cg = _____ g.

   1 g = _____ cg.

112. 1 mg = _____ g.

     1 g = _____ mg.

     4 mg + 3 mg = _____ g.

     4 g + 3 g = _____ mg.

Answers for Programmed Lessons

1. deci-
   centi-
   milli-
2. deka-
   hecto-
   kilo-
3. meter
4. centimeter
   kilometer
5. meters
6. 6m
   no
7. a pencil;
   a delicate instrument,
   e.g., surgical; a transistor
8. decimeter
   tenth
9. 3 dm
10. centimeter
    1/10 or 0.1
    1/100 or 0.01
11. 100
    1/100 or 0.01
12. 7 cm
13. 600
14. 4m
15. millimeter
16. 1 000
    1/1 000 or 0.001
17. 8 mm
18. 3 000 mm
19. 4 m
20. dekameter
21. 10
22. meters
    1/10 or 0.1
23. 6 dam
    5 dam
24. 10
25. 8 hm
26. 100
27. meter
28. 1/100 or 0.01
29. 3 km
30. 100 ×
    meter
31. 10
    10
    100
    1 000
32. 1 000 ×
    a meter
33. 1/1 000 or 0.001
34. 35 km
35. 6 000 m
    4 km 362 m
36. 0.5 m
37. 4.2 m

38. 0.04 m
    0.06 m
39. 9.04 m
    8.27 m
40. 30 sq. m.
    or 30 m^2
    35 m^2 or
    35 sq. meters
41. see page 34
42. 9 m^3
43. 8 m^3
44. 1 decimeter or 1 dm
45. 8 dm^3
46. 9 dm^3
47. 1 centimeter
48. cm^3
49. 5 cm^3
50. 5 cm^3
51. 64 cm^3
52. 27 cm^3
53. 1 000
    1 000
54. 1
    6 1
55. 10 × 10 × 10
56. 1/100 or 0.01
    1/1 000 or 0.001
57. deciliter
    or 0.1 1
58. 5 dl
    7 dl
59. 0.1 1
60. 0.5
    0.4 1
61. 50
62. centiliter
63. centi-
    liter
64. 4 cl
    0.04
65. 0.06 1
66. 4.02 1

67. 7.67 1
68. 100
69. 700
70. 311
71. milliliter
    1/1 000
72. m l
    m milli-
    1 liter
73. 6 ml
    8 ml
74. milli-
    0.001
75. 0.031 1
76. 2.416 1
77. 10
    100
    1 000
78. 10
    100
    1 000
79. 10 ×
    1/100
    1/1 000
80. 6 112 ml
81. kg
    k
    gram
82. 7 kg
83. 4 000
84. 4 600
85. 9 250
86. 100 ×
87. 100
88. hecto- or 100 ×
    g
    hectogram
    100
    kilogram
89. 10 ×
    10
    dekagram

90. deka-
gram
91. dekagram
10 ×
10
92. 7 dag
93. 4 g
94. 1/1 000
95. 0.002 kg
0.025 kg
0.275 kg
96. 3.004
97. deci-
g
decigram
dg
20
98. 60
600
99. 60
600
6 000
100. centi-
gram
1/100
101. centigram

102. 5 cg
103. gram
700
6
104. 1/1 000
1/1 000
1 000
105. milli-
gram
106. 9 mg
107. 2 000
7 000
7
108. 0.001
0.006
109. 0.001
1 000
0.045
45 000
110. 0.1 or 1/10
10
111. 0.01 or 1/100
100
112. 0.001 or 1/1 000
1 000
0.007
7 000

CHAPTER VI

HOW WILL ADOPTION OF THE METRIC SYSTEM AFFECT SPECIFIC ACTIVITIES AND INDIVIDUALS?

Weights and measures may be ranked among the necessaries of life to every individual of human society. They enter into the economical arrangements and daily concerns of every family. They are necessary to every occupation of human industry; to the distribution and security of every species of property; to every transaction of trade and commerce; to the labors of the husbandman; to the ingenuity of the artificer; the studies of the philosopher; to the researches of the antiquarian; to the navigation of the mariner, and the marches of the soldier; to all the exchanges of peace, and all the operations of war. The knowledge of them, as in established use, is among the first elements of education, and is often learned by those who learn nothing else, not even to read and write. This knowledge is riveted in the memory by the habitual application of it to the employments of men throughout life.

—*John Quincy Adams;*
Report to the Congress, 1821

Although the completed U.S. metric study weighed in at only 17 ounces, or more appropriately 481.95 grams, its proposals were of multi-megaton proportions in terms of its effects on Americans. We can expect the conversion of nearly all U.S. weights and measures to the metric system during the next ten years. This metric conversion will require us to think and do in metric terms on our assembly lines, when buying a new shirt or gasoline, or when writing grade-school math books. It will mean that the Statue of Liberty stands at 46.050 meters from base to torch, that 452 200 000 liters of water roar over Niagra Falls every minute, that the Buffalo Bills have the ball and its third down and 8.233 meters to go, that it's 9° Celsius, that you've gained a kilogram and expanded a centimeter or two around the middle, that you'll buy a liter of milk, that water freezes at 0° and boils at 100°, that the hamburger you're about to eat contains 1 024 joules and the piece of apple pie 1 463 joules (or we may just decide that each calorie contains 4 180 joules in order to simplify matters). Think of Peter Piper picking 8.810 liters of pickled peppers, a normal body temperature of 37 degrees Celsius, and a Miss America whose centimetrical measurements are 91.4 cm/61.0 cm/91.4 cm.

The Experience of New Zealand

Let's take a look at another nation which is now facing and solving some of the metric conversion problems that we shall soon be facing and solving—New Zealand.

WELLINGTON, NEW ZEALAND, DECEMBER 6, 1972— Metrication! The word strikes terror in the hearts of grocery clerks, shopkeepers, teachers, carpenters, and real estate agents—not to mention Mr. and Mrs. Average Consumer.

By the same token, metrication enthralls scientists, computer programmers, weathermen, exporters, bankers, and others with an orderly turn of mind.

As this South Pacific nation begins the slow, painful, and often mind-blowing switch-over to metric measurements, the perspiration starting to form on the average brow as Joe Kiwi mumbles to himself, "What is a kilometer, anyway?"

The change from imperial units to metric units is unquestionably a rational move—two-thirds of the world's nations use the metric system these days and most international trade is carried out in metric terms. Metrics has long been the language of scientists, technicians, and mathematicians. The metric system, once learned, is simpler, more precise, and more standardized.

Yet the complications and ramifications of reorganizing an entire nation's measuring system boggle even the metricated mind.

After January, for example, all real estate transactions will be computed metrically. This means that lawyers, realtors, banks, finance companies, sellers, and buyers will suddenly be transported into a misty realm where an acre is no longer an acre but 2.47 hectares, instead.

Says one Wellington lawyer, "Frankly, I haven't the slightest idea how to do it."

In New Zealand, the changeover started subtly about five years ago with the switch from British currency (pounds, shillings, and pence) to the decimal (dollars and cents) system. Then, last year the TV weather reporters began giving temperature readings in Celsius (centigrade) as opposed to the traditional Fahrenheit. So on a scorching summer day in Christchurch, the temperature is read as a mere 30 degrees (Celsius) instead of the 90 degrees (Fahrenheit) it was in the good old days.

Quite simple, explains the announcer; add 15 to the Celsius reading and double it. This comes out as an approximate conversion to Fahrenheit.

No sooner was that fact partially digested than the highway mileage signs underwent a strange metamorphosis. The distance from the one place to another mysteriously lengthened. It was no longer 25 miles but 40 kilometers instead.

From this point on, things get really sticky. Within the next year or so, New Zealanders must prepare to wrestle with metric infiltration into everyday life. A pound of butter will be 453.5 grams; a two-by-four will become a 50.8 mm-by-101.6 mm; a six-foot man will stand 1.82 meters tall; a pint of milk will read out at 568 milliliters.

As this phase occurs, a marked increase in emotional

breakdowns is to be expected. Perhaps the first to go will be the dairymen who have been putting milk in imperial pint bottles since time immemorial: not only will the measurement change, but the bottle size and the price as well.

As the effect snowballs, uncontrollable sobs are likely to be heard from grocery stores (who ever heard of one-half kilogram of carrots?); from manufacturing firms where all equipment will have to be recalibrated to millimeters and centimeters, grams and dekagrams in place of inches, feet, ounces, and pounds; and in primary schoolrooms where teachers will be required to herd their pupils through a maze of millis, centis, decis, and kilos.

In twenty years or so, the experts hope, such terms as "ton" and "bushel," "yard" and "mile," "ounce" and "gallon" will be pleasantly archaic.

In the process, of course, some of the color and sparkle leave the language forever. For example, no one could successfully merchandise a "three-decimeter hot dog" . . . no ardent swain is going to croon, "I love you a hectoliter and a dekaliter" . . . nor will the farmer send his son out to plow the south 16.19 hectares.

However, the swing to metrics is inevitable, since 131 countries currently use the system, and still more are contemplating the switch. Australia is on the bandwagon, about one year ahead of New Zealand. Among the largest laggards are the U.S. and Canada, where metrication is still in the talking stage.

The New Zealand public education program to prepare for the change is a carefully orchestrated one. In general, the approach is low-key. Broadcasting personnel cooperate with informational spots explaining how to convert from imperial to metric in the simplest possible terms. Downtown banks advertise free scales giving your weight metrically. Post offices distribute "Metric Memos" and display posters.

As the deadline draws nearer, there will be a concerted, government-backed effort to retrain workers in the art of counting by tens and instructing them in the somewhat forbidding jargon which goes with metrication. Says one New Zealand specialist in the field, "This is one of the biggest changes we will have taken in our lifetime."

In New Zealand, with a population of slightly less than 3,000,000, total conversion to metrics is expected to require nearly ten years and to price out in the neighborhood of $2,000,000 in administrative costs alone. An effective target date is 1976 for the change-over ... the hangover may take longer.

"Bartender, three centimeters of scotch on ice, please ... and leave the liter."

(This article is reprinted with the permission of the writer, John Forbis of Wellington, New Zealand. It appeared in the *Buffalo Evening News* of December 6, 1972, and in other American newspapers.)

Education

The United States Metric Study reported that educators (and people in related fields), totaling 1 600 000 persons, almost unanimously endorse the metric system. The National Education Association is endorsing a carefully planned, concerted effort to convert to the metric system, which it believes is essential to the future of American industrial and technological development and to the evolution of effective world communication. It urged that American teachers teach metric as the primary system of weights and measures.

The education segment of the U.S. Metric Study reported that the metrication of textbooks and equipment would cost one billion dollars over a three-to-five-year period. However, this cost cannot be attributed to a metric changeover, because textbooks are replaced anyway every three to five years. The billion dollars would not appear as an extra item in school budgets.

The training of teachers into the use of the metric system should be a minor cost, since most teachers participate in continuing education programs. Teacher training into metrication could take place in the form of in-service programs or educational TV series.

Some project the estimate that $700 million a year in

teachers' time may be saved if schoolchildren are no longer required to learn the complex system we now use. In some grades, as much as 25 percent of teaching time is devoted to arithmetic focusing on our customary system of weights and measures. Some observers have reported that small children find it easier to conceptualize centimeters and millimeters than to interpret the graduations on a customary ruler and the small fractions of an inch. A study completed by the American Association for the Advancement of Science concluded that slower children learn metric more readily than they do the customary system. An Australian metric study arrived at the same conclusion. The same unity could be achieved in this area of math education as exists in the ordinary decimal concepts in our extraordinarily simple monetary system.

Metrication will be a joy to schoolchildren, who will save up to 25 percent of their math time and have an easier time of it in terms of difficulty of learning. One particular educator estimated that we may even eliminate two years of elementary-school arithmetic because of the elimination of the teaching of the cumbersome, antique fractions. While endorsing the metric system, the American Geophysical Union chairman remarked that monstrosities such as proper and improper fractions, numerators, least common denominators, greatest common divisors, and mixed numbers could at last be laid to rest with the celluloid collar and the oxcart.

Given a ten-year conversion schedule, the Bureau of Standards, after two years of study and preparation, recommends that SI be taught from kindergarten through the sixth grade as a "first language," and that SI be taught exclusively in grades seven to twelve. The metric (SI) system would be taught exclusively in all elementary and secondary schools after the eighth year of the conversion program.

The U.S. educational establishment will play the lead role in introducing the metric system to new generations of youth. The gradual shift into metrics will include books, maps, materials and equipment, courses, and methods of teaching. Courses in math and science will be affected first, and other courses shall be included later, as even the cooks in the school cafeterias change their recipes from fractions and customary weights to metrics. The goal of school boards, teachers, and administrators is to cause youngsters to think in metric units, to visualize relative sizes and quantities in metric units.

School districts that are introducing the metric system will have to consider the following five stages:

1. Teachers must be thoroughly prepared to teach the metric system. This may require seven to fifteen hours or more in-service training, or a college course designed for teachers may be a requirement. Teaching the metric system should be postponed until this step is completed.
2. Curriculum guides, courses of study, textbooks, and other "software" must be revised or replaced.
3. Classrooms, laboratories, and industrial-arts departments will need the metric measuring tools—meter sticks instead of yardsticks; maps with metric grid lines with distances scaled in meters instead of miles.
4. School purchasing departments will need to learn metric specifications.
5. School cooks as well as home-economics departments may need to adjust to using metric recipes.

Teachers should avoid integrating the teaching of the metric system with the teaching of standard measurements. The metric system should be taught as such without comparisons or conversions related to the standard system of measurement. A master plan which involves all grades may be important to a successful conversion program. It may well turn out that the young learner is able to master the metric system before adults, including teachers.

A possible curriculum guide to grade six

Preschool. If exposed at all to a measurement system at this time, the child should be exposed to the metric system. Television presentations for children as well as three-dimensional toys, along with family support, should suffice during this period.

Kindergarten. During this time, children are led toward the concepts of size comparisons—big-little, tall-short, heavier-lighter. Exposure to meter lengths, decimeter lengths, liter amounts, and gram weights may be appropriate for many children at this time. TV programs of the "Sesame Street" type seem to be successful with these concepts.

First grade. The child can include metrics in his addition and subtraction activities by using a metric number line or meter stick. The relationship between multiples of ten, such as the centimeter and millimeter, may be introduced. If children can learn to recognize basic amounts with coins at this level, they can also learn the basic metric units. If children can handle lessons which develop the concept of ones place, tens place, hundreths place, and thousandths place, then basic metric prefixes should be related to these lessons. Metric lengths and widths should be emphasized during the next three grades.

Second Grade. Basic metric prefixes should be continued on this level along with the relationships of "more than" and "less than" between metric units of measure. Work should continue with divisions and multiples of ten.

Third grade. Multiples of ten up to one hundred should now be conceptualized by the child with relative ease. Addition on the meter stick with decimeters, centimeters, and millimeters can take place. The Celsius thermometer, like the meter stick, is marked from 0 degrees for freezing to 100 degrees for boiling and can be used for additions and subtractions. Many pupils may now be able to begin estimating and computing within the concepts of area and volume. What is the square of something? What is the cube of something? Again the emphasis is on ten. The kilometer may be introduced on this level. Very elementary computations for finding the answers to: How fast? How far? How much? How long? How wide? How hot? How cold? may be introduced to many pupils at this time and continued in fourth, fifth, and sixth grades.

Fourth grade. History, development, and purpose of the metric system may be introduced in the fourth grade and continued in the fifth and sixth grades. Metric addition and subtraction should be continued. Many children may now be ready for metric averages. Metric equivalences should be continued; for example, 86 on the meter stick is equivalent to 86/100. Metric weights should be introduced with elementary computations continuing through the sixth grade. Work with multiples and divisions of ten and their correlations with metric prefixes should continue with the meter, the liter, and the gram; that is, the three prefixes for multiples of ten and the three prefixes for divisions of ten.

Fifth grade. In-depth implementation of metric terms should proceed on this level. Multiplication and division in metric measurement can accompany traditional problems in multiplication and division. Metric measurement should now accompany the teaching of science hand in hand. Decimal equivalents of each measure should be recognized. Explanation of European development and adaptation of the metric system, including the Treaty of the Meter, should now be accompanied by the history of America's own metric movement.

Sixth grade. Pupils on this level should know both European and American historical developments of the metric system as well as its implications for commerce and industry. More sophisticated relationships between metric measurements and science can be recognized and accompanied by more sophisticated computations for science lessons, e.g., metric time, rate, distance, and weight problems.

In summary, an elementary introduction to the metric system should enable the child to:

1. Recognize and accept the need for learning the metric system.
2. Examine the European and American historical development of the metric system.
3. Recognize the metric prefixes and their value in terms of multiples or divisions of ten.
4. Identify the metric units for measuring length, weight, capacity, and temperature. When presented with a variety of objects, the child should be able to choose the appropriate metric unit for measuring each object.
5. Recognize and compute simple equivalencies from one metric unit to another; that is, be able to shift the decimal point while recognizing what is happening in terms of place value to multiples and divisions of ten.
6. Demonstrate successful computation of simple problems using metric units of measure.

Junior high school

Teaching the metric system on this level (or any level) will depend on whether or not the student has had previous exposure to the metric system as well as on the depth of his exposure. If a student were exposed to the metric system for the first time in the seventh grade, the emphasis on his instruction for the next two years would be on the fundamentals. His advancement to more sophisticated levels would depend on his acquisition of appropriate competencies. A student with appropriate competence on this level should be able to find the metric areas of simple geometric figures. He should be able to apply his competence in the area of linear measure to formulas. The relationship between volume and weight may be conceptualized during the eighth grade. Three-dimensional models of geometric figures should be used for visual perception. If metric units are used in formulas, they should be accompanied by explanations of scientific notation.

Senior high school

Much said earlier about the degree of exposure and level of metric conceptualization applies on the high school level. Metric activities, for those who have acquired the appropriate competence, should take place in the science laboratories, the industrial-arts shops, the home-economics areas, the sports arenas, and the mathematics classes. These metric activities should be in the forms of both computations and measurement.

It is assumed that in genuine open-school and non-graded situations levels of appropriate competences will be used in place of grade levels.

These statements and recommendations should be imprinted by educators when teaching (and learning) the metric system within the structured situations of a public school system.

1. Under no circumstances should teachers be asked to teach the metric system unless they are adequately prepared to do so.
2. In-service programs providing for the adequate preparation of teachers should be the responsibility of school administrations in cooperation with colleges and universities.
3. Regular attendance at these programs is the responsibility of the individual teacher. No teacher should be expected or allowed to teach the metric system unless adequately prepared to do so. Unprepared teachers should be deemed unqualified to teach the metric system. Eventually, teachers not prepared to teach the metric system will not be prepared to teach math or science on any level.
4. When teaching the metric system in any grade or competency level, always begin with concrete perception and conceptualization before moving on to the abstract conceptualization and computation.
5. Avoid completely the making of comparisons or conversions with the traditional system of measurement. The student needs to learn the metric system as a primary language separate and apart from the traditional system. Eventually, the learner will make his own comparisons between the liter and the quart and the kilometer and the mile.

Curriculum and instruction

A National Metric Education Center has been funded by the U.S. Office of Education for the purpose of providing course materials and research facilities for teachers. The center will analyze the problems that will be encountered in converting the present system of measurement to the metric system in such areas as construction, electronics, aerospace, and health occupations. The purpose of the project is to prepare teachers to train others in teaching the metric system and the use of the equipment geared to metric measure. The center will cooperate with national associations which are involved with metrication. It will assist both educators and industrial concerns with in-service metric training programs.

During its first year, the center will develop instructional materials and recommend changes in teacher education programs. A model training program and in-service workshops for teachers are on its agenda for the second and third years.

The methods of teaching metric will vary and will equal the number of methods of teaching. The author, who himself is an educator, at this time supports the philosophy of teaching metric that holds that the key to teaching the metric system is to avoid any comparison at all with inches, quarts, pounds, and other elements of the customary system. The state of California will use this philosophy beginning in 1976, when the metric system will be the only standard of weights and measures taught to its 4.5 million schoolchildren. Instruction that converts our present system into metrics and metrics into our present system is to be avoided in California. California math and science textbooks will be written only in metric. An interesting case study of the methods used by one teacher in the elementary school was presented in the education section of the May 7, 1973 *Newsweek*, "Meter Readers." The slogan for math education of the '70s will be: Think Metric!

Sports

Metrication may not be much of a problem in sports. Internationally, soccer is by far the most popular game; there is no standard size for a soccer field. Cricket is played throughout the British Empire; football in the United States. It may be unnecessary to change the lengths of the football fields. However, the Marylebone Cricket Club has organized a metrication subcommittee that has altered the rules to include metric equivalents. A golf club in East Kilbride, Scotland, has placed markers on the tees showing distances in meters only. FINA, the international swimming federation, is refusing to ratify any more records on Imperial measurements. Metric equivalents are being adopted for most British sports.

Farming

Farmers may not have as much of a psychological problem in thinking in a new measurement language because they are so accustomed to mental computations. However, the cost of retooling and replacing machinery will be, in part, passed on to the farmer. Some inconvenience and higher repair costs may be experienced during the changeover on machinery presently owned. It will not make any difference if a farmer orders a new 14-foot or a 4.2-meter cutter bar for a combine head. The engineers at the factory will, however, design headers to measure in whole numbers just as we now do in the customary system. Instead of a 14-foot header, they may design a 3-, 4-, or 5-meter header. The new parts may not fit present machinery. This should not be a problem to farmers. Most, if not all, present machinery may be replaced by metric-sized machinery before this becomes a serious problem.

Farmers will be speaking of corn yields averaging 88.9 quintals per hectare or hogs weighing an average of 100 kilograms. But farmers are accustomed to change as a way of life, because they have had to adapt to change in order to survive in a highly competitive marketplace. Metrication is just another one of such necessary changes for farmers.

Culinary Arts

Recipes will be simplified and the food shopper will be able to compare food prices and package sizes to get the best buys; almost an impossibility at this time. Bothersome fractions will be replaced by a system which facilitates mental computation.

We will need to become accustomed to buying meat by the kilo (kilogram or 1 000 grams), which is 2.2. pounds. A 4- to-6-pound family roast will weigh 2 to 3 kilos, chickens will weigh 1½ kilos, cold cuts will be sold by the 100 grams (3½ ounces), a stick of butter will weigh 120 grams (⅛ kilo). Flour will be measured in grams, the average cup weighing 100 to 120 grams (pastry flour 80 grams per cup).

$$5 \text{ milliliters} = 1 \text{ teaspoon}$$
$$15 \text{ milliliters} = 1 \text{ tablespoon}$$

$$1 \text{ cup} = 250 \text{ milliliters}$$
$$¾ \text{ cup} = 200 \text{ milliliters}$$
$$½ \text{ cup} = 150 \text{ milliliters}$$
$$¼ \text{ cup} = 50 \text{ milliliters}$$

Small Business

Small businesses are being placed at a disadvantage because they lack the technical, financial, and managerial resources for planning and coping with the metric system. Because large companies set the pace, the small businessman is less likely to make his own decision as to when to go metric. The small businessman is more likely to be dependent on the availability of standard parts from suppliers. A coordinated national program will especially favor the small business and self-employed craftsmen. Vocational, on-the-job training programs and technical assistance may be supported by the government.

The Motorist

The sign of the times will be road signs showing distance in kilometers as well as in miles. These signs are already in use in some states. Plans specify for an all-metric America by 1983 accompanied by road signs showing distances in kilometers only. Gasoline will be sold metrically (it is now sold by the liter in Europe) instead of by the gallon. Automobile parts and tools are coming in metric sizes. In essence we will drive metric-sized cars repaired with metric-sized tools, powered by metrically measured fuel on metrically measured roads, following metric road signs and speed limits.

The Consumer

The U.S. Metric Study reports that a comprehensive program of public education will be required for successful conversion to the metric system. Most consumers seem to agree that the advantages will outnumber the inconveniences after metrication is completed. Metrication will foster innovation and invention. Among the possibilities are improved standards for clothing sizes, simplified package and can sizes, an end to confusing practices in consumer-products information, and standardization at the international level. Mass consumer education may be a government undertaking.

Construction

The construction industry is one of the most difficult industries to convert to the metric system. New metric dimensions have to be recognized for all building components; length, mass, and volume as well as all the subunits. However, a note of optimism is that in Britain, the construction industry is leading the nation to metrication.

Engineering Standards

Because precise measurement is a basic activity affecting every aspect of an engineering firm's business, in no other industrial syndrome does metrication present so profound a change. All firms, because of metrication, will have to review the design of their products and decide whether or not to change the entire product or merely alter the components. Metrication gives companies the opportunity to clean house in terms of purchasing policies for materials and components, organization of production (industrial engineering), policy concerning stock control, and marketing practices. Usually metrication results in the elimination of unnecessary sizes. The role of the engineering industries is further magnified by the fact that their client industries will rely on them for their own metrication.

Labor

Labor unions too are concerned with metrication in terms of costs of new tools and retaining of their members. In the final analysis, employers may well pay for all tools and retraining. Self-employed craftsmen, however, may have to pay for their own new tools regardless of a national program. These persons may receive government-supported retraining. There may be some inestimatable loss of time and production over an undetermined period of time because of the lack of familiarity with the new tools. Equitable ways of coping with this problem to everyone's satisfaction will be a concern.

Manufacturing

According to the U.S. Metric Study, 70 percent of American manufacturers favored the use of the metric system because it was in the best interests in the United States to adopt it. Industries are discovering that retraining for using the metric system is not the serious obstacle it was a few years ago. The Council for Technical Examining Bodies is preparing metric programs for trainees and workers in the major industries. Vocational schools and technical institutes are designing their metrics training programs to meet the needs of the industries they serve. Small companies may not have the resources to convert as smoothly as large companies.

The automobile industry is one of the largest users of machine parts. The cycle for a complete changeover in parts from any given model year is twelve years: that is, a 1970 model may not contain any 1958 parts. This cycle could be programmed for metric input so that replacements consisted of parts with metric dimensions. Systems analysis would minimize problems in making interconnections. The Ford Motor Company seems to have taken the lead in the metrication of its industry thus far, as evidenced by the metric-conversion education materials it is producing for its own people. General Motors seems aloof and cool to metrication; one hopes that it is not going by the adage "What's good enough for General Motors is good enough for the United States."

CHAPTER VII

COMPREHENSIVE TABLES FOR EQUIVALENTS AND FOR CONVERSION

Equivalence

The complete list of British Weights and Measures as modified by use in the United States is as follows:

Unit	Domestic	Equivalents Metric
LENGTH		
1 mil	0.001 in.	0.025 4 mm
1 point (typographical)	0.013 837 in.	0.351 5 mm
1 line (button measure)	0.025 in.	0.635 0 mm
1 inch		2.540 0 cm
1 hand	4.0 in.	10.160 0 cm
1 link (surveyor's)	0.01 chn.	20.116 8 cm
1 span	9.0 in.	22.860 0 cm
1 foot	12.0 in.	30.480 0 cm
1 pace	30.0 in.	76.200 0 cm
1 vara	33.33 in.	84.666 7 cm
1 yard	3.0 ft.	0.914 4 m
1 fathom	6.0 ft.	1.828 80 m
1 rod, perch or pole	5.5 yd.	5.029 20 m
1 chain (surveyor's)	4.0 rd.	20.116 8 m
1 furlong	220 yd	201.168 m

Unit	Domestic	Equivalents Metric
1 cable	120.0 fathoms	219.456 m
1 mile (statute)	8.0 furlongs	1 609.344 m
1 nautical mile (int.)		1 852.000 m
1 league	3.0 st.mi.	4.828 03 km

AREA

1 sq. inch		6.451 60 cm²
1 sq. link	62.726 4 sq. in.	404.686 cm²
1 sq. foot	144 sq. in.	9.290 30 dm²
1 sq. yard	9 sq. ft.	0.836 127 m²
1 sq. rod	30.25 sq. yd.	25.292 85 m²
1 sq. chain	4 356.0 sq. ft.	4.046 86 a
1 acre	43 560.0 sq. ft.	40.468 6 a
1 labor (Texas)	177.1 ac.	71.670 ha
1 section	1.0 sq. mi.	2.590 km²
1 township	36 sections	93.240 km²

VOLUME

1 cubic inch		16.387 1 cm³
1 cubic foot	1 728 cu. in.	28.316 8 dm³
1 cubic yard	27 cu. ft.	0.764 555 m³
1 cord	128 cu. ft.	3.624 56 m³

CAPACITY—DRY

1 dry pint		0.550 61 l
1 dry quart	2 pints	1.101 221 l
1 peck	8 quarts	8.809 77 l
1 bushel	4 pecks	35.239 1 l
1 barrel (cranberries)	5 826 cu. in.	95.471 l
1 barrel (fruits & veg.)	7 056 cu. in.	115.627 l

CAPACITY—LIQUID

1 minim		0.061 61 ml
1 fluid dram	60 minims	3.696 69 ml
1 fluid ounce	8 fl. drams	2.957 35 cl
1 gill	4 fl. ounces	11.829 4 cl

Unit	Domestic	Equivalents Metric
1 liquid pint	4 gills	4.731 76 dl
1 liquid quart	2 liq. pints	9.463 53 dl
1 gallon	8 liq. pints	3.785 41 l
1 barrel (petroleum)	42 gallons	158.987 l

WEIGHT

Avoirdupois measures

Unit	Domestic	Equivalents Metric
1 grain		64.798 9 mg
1 av. ounce	437.5 grains	28.349 5 g
1 av. pound	16 av. oz.	453.592 g
1 short hundredweight	100 av. lb.	45.359 2 kg
1 short ton	2 000 av. lb.	907.185 kg
1 long ton	2 240 av. lb.	1 016.047 kg

APOTHECARIES' OR TROY MEASURES

Unit	Domestic	Equivalents Metric
1 grain		64.798 9 mg
1 scruple	20 grains	1.295 98 g
1 pennyweight	24 grains	1.555 17 g
1 dram	3 scruples	3.887 93 g
1 ounce	8 drams	31.103 5 g
1 pound	12 ounces	373.242 g

Approximate Conversion of Common Units

U.S. to Metric

LENGTH
1 inch	=	25.0 millimeters (mm)
1 foot	=	0.3 meter (m)
1 yard	=	0.9 meter
1 mile	=	1.6 kilometers (km)

AREA
1 sq. inch	=	6.5 sq. centimeters (cm²)
1 sq. foot	=	0.09 sq. meter (m²)
1 sq. yard	=	0.8 sq. meter
1 acre	=	0.4 hectare*
1 sq. mile	=	2.6 sq. kilometers

MASS
1 grain	=	64.8 milligrams (mg)
1 ounce (dry)	=	28.0 grams (g)
1 pound	=	0.45 kilogram (kg)
1 short ton	=	9.071 kilograms

Metric to U.S.

LENGTH
1 millimeter (mm)	=	0.04 inch
1 meter (m)	=	3.3 feet
1 meter	=	1.1 yards
1 kilometer (km)	=	0.6 mile

AREA
1 sq. centimeter (cm²)	=	0.16 sq. inch
1 sq. meter (m²)	=	11.0 sq. feet
1 sq. meter	=	1.2 sq. yards
1 hectare	=	2.5 acres
1 sq. kilometer	=	0.39 sq. mile

MASS
1 milligram (mg)	=	0.015 grain
1 gram (g)	=	0.035 ounce
1 kilogram (kg)	=	2.2 pounds
1 metric ton	=	1.102 tons (short)

*1 hectare equals 10,000 sq. meters.

U.S. to Metric

VOLUME

1 cubic inch	= 16.0 cubic centimeters (cm³)
1 cubic foot	= 0.03 cubic meter (m³)
1 cubic yard	= 0.76 cubic meter
1 teaspoon	= 5.0 milliliters (ml)
1 tablespoon	= 15.0 milliliters
1 fl. ounce	= 30.0 milliliters
1 cup	= 0.24 liter (1) *
1 pint	= 0.47 liter
1 quart (liq.)	= 0.95 liter
1 gallon (liq.)	= 0.004 cubic meter
1 peck	= 0.009 cubic meter
1 bushel	= 0.04 cubic meter

POWER

1 horsepower	= 0.75 kilowatt (kw)

ENERGY

1 calorie	= 4.18 joules (j)

* 1 liter equals 1 cubic decimeter (dm³).

Metric to U.S.

VOLUME

1 cubic centimeter (cm³)	= 0.06 cubic inch
1 cubic meter (m³)	= 35.0 cubic feet
1 cubic meter	= 1.3 cubic yards
1 milliliter (ml)	= 0.2 teaspoon
1 milliliter	= 0.07 tablespoon
1 milliliter	= 0.03 ounce
1 liter (1)	= 4.2 cups
1 liter	= 2.1 pints
1 liter	= 1.1 quarts
1 cubic meter	= 264.0 gallons
1 cubic meter	= 113.0 pecks
1 cubic meter	= 28.0 bushels

POWER

1 kilowatt (kw)	= 1.3 horsepower

ENERGY

1 joule (j)	= 0.24 calorie

International Nautical Miles and Kilometers

Miles to Kilometers

International Nautical Miles	Kilometers	International Nautical Miles	Kilometers
0		50	92.600
1	1.852	51	94.452
2	3.704	52	96.304
3	5.556	53	98.156
4	7.408	54	100.008
5	9.260	55	101.860
6	11.112	56	103.712
7	12.964	57	105.564
8	14.816	58	107.416
9	16.668	59	109.268
10	18.520	60	111.120
11	20.372	61	112.972
12	22.224	62	114.824
13	24.076	63	116.676
14	25.928	64	118.528
15	27.780	65	120.380
16	29.632	66	122.232

Kilometers to Miles

Kilometers	International Nautical Miles	Kilometers	International Nautical Miles
0		50	26.997 8
1	0.540 0	51	27.537 8
2	1.079 9	52	28.077 8
3	1.619 9	53	28.617 7
4	2.159 8	54	29.157 7
5	2.699 8	55	29.697 6
6	3.239 7	56	30.237 6
7	3.779 7	57	30.777 5
8	4.319 7	58	31.317 5
9	4.859 6	59	31.857 5
10	5.399 6	60	32.397 4
11	5.939 5	61	32.937 4
12	6.479 5	62	33.477 3
13	7.019 4	63	34.017 3
14	7.559 4	64	34.557 2
15	8.099 4	65	35.097 2
16	8.639 3	66	35.637 1

Miles to Kilometers

International Nautical Miles	Kilometers	International Nautical Miles	Kilometers
17	31.484	67	124.084
18	33.336	68	125.936
19	35.188	69	127.788
20	37.040	70	129.640
21	38.892	71	131.492
22	40.744	72	133.344
23	42.596	73	135.196
24	44.448	74	137.048
25	46.300	75	138.900
26	48.152	76	140.752
27	50.004	77	142.604
28	51.856	78	144.456
29	53.708	79	146.308
30	55.560	80	148.160
31	57.412	81	150.012
32	59.264	82	151.864
33	61.116	83	153.716
34	62.968	84	155.568
35	64.820	85	157.420

Kilometers to Miles

Kilometers	International Nautical Miles	Kilometers	International Nautical Miles
17	9.179 3	67	36.177 1
18	9.719 2	68	36.717 1
19	10.259 2	69	37.257 0
20	10.799 1	70	37.797 0
21	11.339 1	71	38.336 9
22	11.879 0	72	38.876 9
23	12.419 0	73	39.416 8
24	12.959 0	74	39.956 8
25	13.498 9	75	40.496 8
26	14.038 9	76	41.036 7
27	14.578 8	77	41.576 7
28	15.118 8	78	42.116 6
29	15.658 7	79	42.656 6
30	16.198 7	80	43.196 5
31	16.738 7	81	43.736 5
32	17.278 6	82	44.276 5
33	17.818 6	83	44.816 4
34	18.358 5	84	45.356 4
35	18.898 5	85	45.896 3

Miles to Kilometers

International Nautical Miles	Kilometers	International Nautical Miles	Kilometers
36	66.672	86	159.272
37	68.524	87	161.124
38	70.376	88	162.976
39	72.228	89	164.828
40	74.080	90	166.680
41	75.932	91	168.532
42	77.784	92	170.384
43	79.636	93	172.236
44	81.488	94	174.088
45	83.340	95	175.940
46	85.192	96	177.792
47	87.044	97	179.644
48	88.896	98	181.496
49	90.748	99	183.348
		100	185.200

Kilometers to Miles

Kilometers	International Nautical Miles	Kilometers	International Nautical Miles
36	19.438 4	86	46.436 3
37	19.978 4	87	46.976 2
38	20.518 4	88	47.516 2
39	21.058 3	89	48.056 2
40	21.598 3	90	48.596 1
41	22.138 2	91	49.136 1
42	22.678 2	92	49.676 0
43	23.218 1	93	50.216 0
44	23.758 1	94	50.755 9
45	24.298 1	95	51.295 9
46	24.838 0	96	51.835 9
47	25.378 0	97	52.375 8
48	25.917 9	98	52.915 8
49	26.457 9	99	53.455 7
		100	53.995 7

Equivalents of Decimal and Binary Fractions of an Inch Expressed in Millimeters

From 1/64 to 1 inch

1/2's	1/4's	8ths	16ths	32ds	64ths	Milli-meters	Decimals of an Inch
					1	0.397	.015 625
				1	2	.794	.031 25
					3	1.191	.046 875
			1	2	4	1.588	.062 5
					5	1.984	.078 125
				3	6	2.381	.093 75
					7	2.778	.109 375
		1	2	4	8	3.175	.125 0
					9	3.572	.140 625
				5	10	3.969	.156 25
					11	4.366	.171 875
			3	6	12	4.762	.187 5
					13	5.159	.203 125
				7	14	5.556	.218 75
					15	5.953	.234 375
	1	2	4	8	16	6.350	.250 0
					17	6.747	.265 625
				9	18	7.144	.281 25
					19	7.541	.296 875
			5	10	20	7.938	.312 5
					21	8.334	.328 125
				11	22	8.731	.343 75
					23	9.128	.359 375
		3	6	12	24	9.525	.375 0
					25	9.922	.390 625
				13	26	10.319	.406 25
					27	10.716	.421 875
			7	14	28	11.112	.437 5
					29	11.509	.453 125
				15	30	11.906	.468 75

Fractions of an inch, with millimeter and decimal equivalents (64ths 31–47)

1/2's	1/4's	8ths	16ths	32ds	64ths	Millimeters	Decimals of an Inch
					31	12.303	.484 375
1	2	4	8	16	32	12.700	.5
					33	13.097	0.515 625
				17	34	13.494	.531 25
					35	13.891	.546 875
			9	18	36	14.288	.562 5
					37	14.684	.578 125
				19	38	15.081	.593 75
					39	15.478	.609 375
		5	10	20	40	15.875	.625
					41	16.272	.640 625
				21	42	16.669	.656 25
					43	17.066	.671 875
			11	22	44	17.462	.687 5
					45	17.859	.703 125
				23	46	18.256	.718 75
					47	18.653	.734 375

Fractions of an inch, with millimeter and decimal equivalents (64ths 48–64)

(whole)	1/2's	1/4's	8ths	16ths	32ds	64ths	Millimeters	Decimals of an Inch
		3	6	12	24	48	19.050	.75
						49	19.447	.765 625
					25	50	19.844	.781 25
						51	20.241	.796 875
				13	26	52	20.638	.812 5
						53	21.034	.828 125
					27	54	21.431	.843 75
						55	21.828	.859 375
			7	14	28	56	22.225	.875
						57	22.622	.890 625
					29	58	23.019	.906 25
						59	23.416	.921 875
				15	30	60	23.812	.937 5
						61	24.209	.953 125
					31	62	24.606	.968 75
						63	25.003	.984 375
1	2	4	8	16	32	64	25.400	1.000

U.S. Equivalents of Common Engineering Units

Length

Kilometer (km)	=	0.621 371 mile
Meter (m)	=	1.093 61 yards
Millimeter (mm)	=	0.039 370 1
Micrometer (μm)	=	39.370 1 microinches

Area

Square kilometer (km²)	=	247.105 acres
Square meter (m²)	=	1.195 99 square yards
Square millimeter (mm²)	=	0.001 550 00 square inch

Volume

Cubic meter (m³)	=	1.307 95 cubic yards
Cubic decimeter (dm³)	=	0.035 314 7 cubic foot
Cubic centimeter (cm³)	=	0.061 023 7 cubic inch
Liter (l)	=	0.264 2 gallon (U.S.)

Velocity

Kilometer per hour (km/h)	=	.621 371 mile per hour
Meter per second (m/s)	=	3.280 84 feet per second

Acceleration

Meter per second per second (m/s²)	=	3.280 84 feet per second per second

Mass

Kilogram (kg)	=	2.204 62 pounds
Gram (g)	=	0.035 274 0 ounce

Density

Kilogram per cubic meter (kg/m³)	=	0.062 428 0 pound per cubic foot

Force

Newton (N) = 0.224 809 pound-force

Torque

Newton meter (Nm) = 0.737 562 pound-force foot

Pressure, Stress

Newton per square meter (N/m²) = 0.000 145 038 pound-force per square inch

Viscosity (Dynamic)

Newton second per square meter
  (Ns/m²)   = 0.020 885 4 pound-force second per square foot

Viscosity (Kinematic)

Square meter per second (m²/s) = 10.763 9 square feet per second

Energy

Joule (j)   = 0.737 562 foot pound-force
Kilojoule (kj) = 0.277 778 watt-hour

Conversion Factors

Units of Length

To Convert from Centimeters

To	*Multiply by*
Inches	0.393 700 8
Feet	0.032 808 40
Yards	0.010 936 13
Meters	0.01

To Convert from
Meters

To	*Multiply by*
Inches	39.370 08
Feet	3.280 840
Yards	1.093 613
Miles	0.000 621 37
Millimeters	1 000
Centimeters	100
Kilometers	0.001

To Convert from
Inches

To	*Multiply by*
Feet	0.083 333 33
Yards	0.027 777 78
Centimeters	2.54
Meters	0.025 4
Inches	12

To Convert from
Feet

To	*Multiply by*
Yards	0.333 333 3
Miles	0.000 189 39
Centimeters	30.48
Meters	0.304 8
Kilometers	0.000 304 8

To Convert from
Yards

To	*Multiply by*
Inches	36
Feet	3
Miles	0.000 568 18
Centimeters	91.44
Meters	0.914 4

To Convert from
Miles

To	Multiply by
Inches	63 360
Feet	5 280
Yards	1 760
Centimeters	160 934.4
Meters	1 609.344
Kilometers	1.609 344

Units of Mass

To Convert from
Grams

To	Multiply by
Grains	15.432 36
Avoirdupois drams	0.564 383 4
Avoirdupois ounces	0.035 273 96
Troy ounces	0.032 150 75
Troy pounds	0.002 679 23
Avoirdupois pounds	0.002 204 62
Milligrams	1 000
Kilograms	0.001

To Convert from
Kilograms

To	Multiply by
Grains	15 432.36
Avoirdupois drams	564.383 4
Avoirdupois ounces	35.273 96
Troy ounces	32.150 75
Troy pounds	2.679 229
Avoirdupois pounds	2.204 623
Grams	1 000
Short hundredweights	0.022 046 23

Short tons	0.001 102 31
Long tons	0.000 984 2
Metric tons	0.001

To Convert from
Metric Tons

To	Multiply by
Avoirdupois pounds	2 204.623
Short hundredweights	22.046 23
Short tons	1.102 311 3
Long tons	0.984 206 5
Kilograms	1 000

To Convert from
Grains

To	Multiply by
Avoirdupois drams	0.036 571 43
Avoirdupois ounces	0.002 285 71
Troy ounces	0.002 083 33
Troy pounds	0.000 173 61
Avoirdupois pounds	0.000 142 86
Milligrams	64.798 91
Grams	0.064 798 91
Kilograms	0.000 064 798 91

To Convert from
Avoirdupois Ounces

To	Multiply by
Grains	437.5
Avoirdupois drams	16
Troy ounces	0.911 458 3
Troy pounds	0.075 954 86
Avoirdupois pounds	0.062 5
Grams	28.349 523 125
Kilograms	0.028 349 523 125

To Convert from
Short Hundredweights

To	Multiply by
Avoirdupois pounds	100
Short tons	0.05
Long tons	0.044 642 86
Kilograms	45.359 237
Metric tons	0.045 359 237

To Convert from
Avoirdupois Pounds

To	Multiply by
Grains	7 000
Avoirdupois drams	256
Avoirdupois ounces	16
Troy ounces	14.583 33
Troy pounds	1.215 278
Grams	453.592 37
Kilograms	0.453 592 37
Short hundredweights	0.01
Short tons	0.000 5
Long tons	0.000 446 428 6
Metric tons	0.000 453 592 37

To Convert from
Short Tons

To	Multiply by
Avoirdupois pounds	2 000
Short hundredweights	20
Long tons	0.892 857 1
Kilograms	907.184 74
Metric tons	0.907 184 74

To Convert from
Long Tons

To	Multiply by
Avoirdupois ounces	35 840
Avoirdupois pounds	2 240
Short hundredweights	22.4
Short tons	1.12
Kilograms	1 016.046 908 8
Metric tons	1.016 046 908 8

To Convert from
Troy Ounces

To	Multiply by
Grains	480
Avoirdupois drams	17.554 29
Avoirdupois ounces	1.097 143
Troy pounds	0.083 333 3
Avoirdupois pounds	0.068 571 43
Grams	31.103 476 8

To Convert from
Troy Pounds

To	Multiply by
Grains	5 760
Avoirdupois drams	210.651 4
Avoirdupois ounces	13.165 71
Troy ounces	12
Avoirdupois pounds	0.822 857 1
Grams	373.241 721 6

Units of Capacity, Volume, and Liquids

To Convert from Milliliters

To	Multiply by
Minims	16.230 73
Liquid ounces	0.033 814 02
Gills	0.008 453 5
Liquid pints	0.002 113 4
Liquid quarts	0.001 056 7
Gallons	0.000 264 17
Cubic inches	0.061 023 74
Liters	0.001

To Convert from Liters

To	Multiply by
Liquid ounces	33.814 02
Gills	8.453 506
Liquid pints	2.113 376
Liquid quarts	1.056 688
Gallons	0.264 172 05
Cubic inches	61.023 74
Cubic feet	0.035 314 67
Milliliters	1 000
Cubic meters	0.001
Cubic yards	0.001 307 95

To Convert from Cubic Meters

To	Multiply by
Gallons	264.172 05
Cubic inches	61 023.74
Cubic feet	35.314 67
Liters	1 000
Cubic yards	1.307 950 6

To Convert from
Minims

To	Multiply by
Liquid ounces	0.002 083 33
Gills	0.000 520 83
Cubic inches	0.003 759 77
Milliliters	0.061 611 52

To Convert from
Gills

To	Multiply by
Minims	1 920
Liquid ounces	4
Liquid pints	0.25
Liquid quarts	0.125
Gallons	0.031 25
Cubic inches	7.218 75
Cubic feet	0.004 177 517
Milliliters	118.294 118 25
Liters	0.118 294 118 25

To Convert from
Liquid Pints

To	Multiply by
Minims	7 680
Liquid ounces	16
Gills	4
Liquid quarts	0.5
Gallons	0.125
Cubic inches	28.875
Cubic feet	0.016 710 07
Milliliters	473.176 473
Liters	0.473 176 473

To Convert from
Liquid Ounces

To	*Multiply by*
Minims	480
Gills	0.25
Liquid pints	0.062 5
Liquid quarts	0.031 25
Gallons	0.007 812 5
Cubic inches	1.804 687 5
Cubic feet	0.001 044 38
Milliliters	29.573 53
Liters	0.029 573 53

To Convert from
Cubic Feet

To	*Multiply by*
Liquid ounces	957.506 5
Gills	239.376 6
Liquid pints	59.844 16
Liquid quarts	29.922 08
Gallons	7.480 519
Cubic inches	1 728
Liters	28.316 846 592
Cubic meters	0.028 316 846 592
Cubic yards	0.037 037 04

To Convert from
Cubic Inches

To	*Multiply by*
Minims	265.974 0
Liquid ounces	0.554 112 6
Gills	0.138 528 1
Liquid pints	0.034 632 03
Liquid quarts	0.017 316 02
Gallons	0.004 329 0

Cubic feet	0.000 578 7
Milliliters	16.387 064
Liters	0.016 387 064
Cubic meters	0.000 016 387 064
Cubic yards	0.000 021 43

To Convert from
Cubic Yards

To	Multiply by
Gallons	201.974 0
Cubic inches	46 656
Cubic feet	27
Liters	764.554 857 984
Cubic meters	0.764 554 857 984

To Convert from
Liquid Quarts

To	Multiply by
Minims	15 360
Liquid ounces	32
Gills	8
Liquid pints	2
Gallons	0.25
Cubic inches	57.75
Cubic feet	0.033 420 14
Milliliters	946.352 946
Liters	0.946 352 946

To Convert from
Gallons

To	Multiply by
Minims	61 440
Liquid ounces	128
Gills	32
Liquid pints	8

Liquid quarts	4
Cubic inches	231
Cubic feet	0.133 680 6
Milliliters	3 785.411 784
Liters	3.785 411 784
Cubic meters	0.003 785 411 784
Cubic yards	0.004 951 13

Units of Capacity, or Volume, Dry Measure

To Convert from Liters

To	Multiply by
Dry pints	1.816 166
Dry quarts	0.908 082 98
Pecks	0.113 510 4
Bushels	0.028 377 59
Dekaliters	0.1

To Convert from Dekaliters

To	Multiply by
Dry pints	18.161 66
Dry quarts	9.080 829 8
Pecks	1.135 104
Bushels	0.283 775 9
Cubic inches	610.237 4
Cubic feet	0.353 146 7
Liters	10

To Convert from Cubic Meters

To	Multiply by
Pecks	113.510 4
Bushels	28.377 59

To Convert from
Dry Pints

To	Multiply by
Dry quarts	0.5
Pecks	0.062 5
Bushels	0.015 625
Cubic inches	33.600 312 5
Cubic feet	0.019 444 63
Liters	0.550 610 47
Dekaliters	0.055 061 05

To Convert from
Dry Quarts

To	Multiply by
Dry pints	2
Pecks	0.125
Bushels	0.031 25
Cubic inches	67.200 625
Cubic feet	0.038 889 25
Liters	1.101 221
Dekaliters	0.110 122 1

To Convert from
Pecks

To	Multiply by
Dry pints	16
Dry quarts	8
Bushels	0.25
Cubic inches	537.605
Cubic feet	0.311 114
Liters	8.809 767 5
Dekaliters	0.880 976 75
Cubic meters	0.008 809 77
Cubic yards	0.011 522 74

To Convert from
Bushels

To	Multiply by
Dry pints	64
Dry quarts	32
Pecks	4
Cubic inches	2 150.42
Cubic feet	1.244 456
Liters	35.239 07
Dekaliters	3.523 907
Cubic meters	0.035 239 07
Cubic yards	0.046 090 96

To Convert from
Cubic Inches

To	Multiply by
Dry pints	0.029 761 6
Dry quarts	0.014 880 8
Pecks	0.001 860 10
Bushels	0.000 465 025

To Convert from
Cubic Feet

To	Multiply by
Dry pints	51.428 09
Dry quarts	25.714 05
Pecks	3.214 256
Bushels	0.803 563 95

To Convert from
Cubic Yards

To	Multiply by
Pecks	86.784 91
Bushels	21.696 227

Units of Area

To Convert from
Square Centimeters

To	*Multiply by*
Square inches	0.155 000 3
Square feet	0.001 076 39
Square yards	0.000 119 599
Square meters	0.000 1

To Convert from
Hectares

To	*Multiply by*
Square feet	107 639.1
Square yards	11 959.90
Acres	2.471 054
Square miles	0.003 861 02
Square meters	10 000

To Convert from
Square Feet

To	*Multiply by*
Square inches	144
Square yards	0.111 111 1
Acres	0.000 022 957
Square centimeters	929.030 4
Square meters	0.092 903 04

To Convert from
Acres

To	*Multiply by*
Square feet	43 560
Square yards	4 840
Square miles	0.001 562 5
Square meters	4 046.856 422 4
Hectares	0.404 685 642 24

To Convert from
Square Meters

To	Multiply by
Square inches	1 550.003
Square feet	10.763 91
Square yards	1.195 990
Acres	0.000 247 105
Square centimeters	10 000
Hectares	0.000 1

To Convert from
Square Inches

To	Multiply by
Square feet	0.006 944 44
Square yards	0.000 771 605
Square centimeters	6.451 6
Square meters	0.000 645 16

To Convert from
Square Yards

To	Multiply by
Square inches	1 296
Square feet	9
Acres	0.000 206 611 6
Square miles	0.000 000 322 830 6
Square centimeters	8 361.273 6
Square meters	0.836 127 36
Hectares	0.000 083 612 736

To Convert from
Square Miles

To	Multiply by
Square feet	27 878 400
Square yards	3 097 600
Acres	640
Square meters	2 589 988.110 336
Hectares	258.998 811 033 6

Conversion of Derived Units

To obtain	Multiply	By
Acres	Hectares	2.471
Atmospheres	Feet of water	0.029 50
Atmospheres	Inches of mercury	0.033 42
Atmospheres	Inches of water	0.002 458
Atmospheres	Kgs./sq. meter	9.678×10^{-5}
Atmospheres	Pounds/sq. inch	0.068 04
Board feet	Lumber width (in.) $\times$ $\dfrac{\text{Thickness (in.)}}{12}$	Length (ft.)
British Thermal Units	Foot-pounds	1.286×10^{-3}
B.T.U.	Horsepower-hours	2547
B.T.U.	Kilograms-calories	3.968
B.T.U.	Kilowatt-hours	3415
B.T.U./hr.	Horsepower (boiler)	33 479
B.T.U./min.	Foot-pounds/min.	1.286×10^{-3}
B.T.U./min.	Foot-pounds/sec.	7.717×10^{-2}
B.T.U./min.	Horsepower	42.44
B.T.U./min.	Kilowatts	56.92
Centimeters	Inches	2.540
Centimeters	Kilometers	10^5
Centimeters	Meters	100
Centimeters	Miles	0.1
Centimeters	Millimeters	1.609×10^5
Centimeters/sec.	Feet/min.	0.508 0
Centimeters/sec.	Feet/sec.	30.48
Centimeters/sec.	Kilometers/hr.	27.78
Centimeters/sec.	Meters/min.	1.667
Centimeters/sec.	Miles/min.	2 682
Centimeters/sec.	Miles/hr.	44.70
Cms./sec./sec.	Feet/sec./sec.	30.48
Cms./sec./sec.	Kms./hr./sec.	27.78

To obtain	Multiply	By
Cubic centimeters	Gallons	3 785
Cubic centimeters	Liters	10^3
Cubic feet	Gallons	0.133 7
Cubic feet	Liters	0.035 31
Cubic feet	Pounds of water	0.016 02
Cu. ft./hr.	Gallons/min.	8.020 8
Cubic ft./min.	Miner's inches	1.5
Cubic feet/sec.	Gallons/min.	2.228×10^{-8}
Cubic ft./sec.	Liters/min.	5.886×10^{-4}
Cubic ft./sec.	Million gals./day	1.547 23
Cubic ft./sec.	Pounds of water/min.	2.670×10^{-4}
Cubic inches	Gallons	231
Cubic inches	Liters	61.02
Cubic inches	Ounces (fluid)	1.805
Cubic inches	Pounds of water	27.68
Cubic inches	Quarts (dry)	67.20
Cubic inches	Quarts (liq.)	57.75
Cubic meters	Gallons	3.785×10^{-3}
Cubic meters	Liters	10^{-3}
Cubic yards	Gallons	4.951×10^{-3}
Cubic yards	Liters	1.308×10^{-3}
Degrees	Quadrants (angle)	90
Degrees	Radians	57.30
Degrees	Revolutions	360
Degrees/sec.	Radians/sec.	57.30
Degrees/sec.	Revolutions/min.	6
Degrees/sec.	Revolutions/sec.	360
Drams	Ounces	16
Drams	Pounds	256
Dynes	Grams	980.7
Dynes	Kilograms	980 665
Feet	Kilometers	3 281
Feet	Meters	3.281
Feet	Miles	5 280
Feet/min.	Kilometers/hr.	54.68
Feet/min.	Meters/min.	3.281
Feet/min.	Meters/sec.	196.8
Feet/min.	Miles/hr.	88

To obtain	Multiply	By
Feet of water	Inches of mercury	1.133
Feet of water	Kgs./sq. meter	3.281×10^{-8}
Feet of water	Pounds/sq. foot	0.016 02
Feet of water	Pounds/sq. inch	2.307
Feet/sec.	Feet min.	0.016 67
Feet/sec.	Kilometers/hr.	0.911 3
Feet/sec.	Meters/min.	0.054 68
Feet/sec.	Meters/sec.	3.281
Feet/sec.	Miles/hr.	1.467
Feet/sec.	Miles/min.	88
(ft./hr.)		Area (sq. ft.)
Foot-lbs.	Horsepower-hours	1.98×10^{6}
Foot-pounds	Kilograms-calories	3 086
Foot-lbs.	Kilowatt-hours	2.655×10^{6}
Foot-lbs./min.	Horsepower	33 000
Foot-lbs./min.	Kilowatts	4.425×10^{4}
Foot/pounds/sec.	Foot-pounds/min.	0.016 67
Foot-lbs./sec.	Horsepower	550
Foot-pounds/sec.	Kilograms-cal./min.	51.43
Foot-lbs./sec.	Kilowatts	737.6
Ft./sec./sec.	Kms./hr./sec.	0.911 3
Gallons	Liters	0.264 2
Gallons	Pounds of water	0.119 8
Gals./min.	Overflow rate (ft./hr.)	$0.124 68 \times$ area (sq. ft.)
Gals./sec.	Liters/min.	4.403×10^{-3}
Grams	Grains (troy)	0.064 80
Grams	Hectograms	100
Grams	Kilograms	10^{3}
Grams	Milligrams	10^{-3}
Grams	Ounces	28.349 527
Grams	Ounces (troy)	31.103 481
Grams	Pennyweights (troy)	1.555 17
Grams	Pounds	453.592 4
Grams	Pounds (troy)	373.241 77
Grams/cm.	Pounds/inch	178.6
Grams/cubic cm.	Pounds/cubic foot	0.016 02
Grams/cubic cm.	Pounds/cubic inch	27.68

To obtain	Multiply	By
Grains	Grams	15.43
Grains	Ounces	437.5
Grains	Ounces (troy)	480
Grains	Pennyweights (troy)	24
Grains	Pounds	7 000
Grains	Pounds (troy)	5 760
Grains (avoir.)	Grains (troy)	L
Grains/gal.	Grams/liter	58.417
Grains/Imp. gal.	Parts/million	0.070 16
Grains/U.S. gal.	Parts/million	0.058 4
Horsepower	Foot-pounds/min.	3.030×10^{-5}
Horsepower	Foot-pounds/sec.	1.818×10^{-3}
Horsepower	Kilogram-cal./min.	0.093 51
Horsepower	Kilowatts	1.341
Horsepower (metric)	Horsepower	1.014
Horsepower-hrs.	Foot-pounds	5.050×10^{-7}
Horsepower-hrs.	Kilograms-calories	1.558×10^{-3}
Horsepower-hours	Kilowatt-hours	1.341
Imperial Gallons	Gallons-U.S.	0.832 67
Inches	Meters	39.37
Inches	Millimeters	0.039 37
Inches of mercury	Feet of water	0.882 6
Inches of mercury	Inches of water	0.073 55
Inches of mercury	Kgs./sq. meter	2.896×10^{-3}
Inches of mercury	Pounds/sq. inch	2.036
Kilograms	Grams	10^{-3}
Kilograms	Milliers	10^{3}
Kilogram-calories	Foot-pounds	3.241×10^{-4}
Kilogram-calories	Horsepower-hours	641.7
Kilogram-calories	Kilowatt-hours	860.5
Kg.-calories/min.	Foot-pounds/min.	3.241×10^{-4}
Kg.-calories/min.	Foot-pounds/sec.	1.945×10^{-2}
Kg.-calories/min.	Horsepower	10.70
Kg.-calories/min.	Kilowatts	14.34
Kilogram-meters	Foot-pounds	0.138 3
Kilogram-meters	Horsepower-hours	2.737×10^{5}
Kilogram-meters	Kilowatt-hours	3.67×10^{5}

To obtain	Multiply	By
Kgs./meter	Pounds/foot	1.488
Kgs./cubic meters	Pounds/cubic foot	16.02
Kgs./cubic meter	Pounds/cubic inch	2.768×10^4
Kgs./sq. meter	Feet of water	304.8
Kgs./sq. meter	Inches of water	25.40
Kgs./sq. meter	Inches of mercury	345.3
Kgs./sq. meter	Kgs./sq. millimeter	10^6
Kgs./sq. meter	Pounds/sq. foot	4.883
Kgs./sq. meter	Pounds/sq. inch	703.1
Kilometers	Meters	10^{-3}
Kilometers	Miles	1.609
Kilometers/hr.	Feet/min.	0.018 29
Kilometers/hr.	Feet/sec.	1.097
Kilometers/hr.	Meters/min.	0.06
Kilometers/hr.	Meters/sec.	3.6
Kilometers/hr.	Miles/hr.	1.609
Kilometers/min.	Miles/min.	1.609
Kilometers/min.	Meters/sec.	0.06
Kilowatts	Foot-pounds/min.	2.260×10^{-5}
Kilowatts	Foot-pounds/sec.	1.356×10^{-3}
Kilowatts	Horsepower	0.745 7
Kilowatts	Horsepower (boiler)	9.803
Kilowatts	Kilogram-cal./min.	0.069 72
Kilowatt-hrs.	Foot-pounds	3.766×10^{-7}
Kilowatt-hours	Horsepower-hours	0.745 7
Kilowatt-hours	Kilograms-calories	1.162×10^{-3}
Knots	Feet/sec.	0.592 1
Knots	Kilometers/hr.	0.539 6
Knots	Miles/hr.	0.868 4
Liters	Gallons	3.785
Liters	Hectoliters	100
Liters	Kiloliters	10^3
Liters	Ounces (fluid)	0.029 57
Liters	Milliliters	10^{-3}
Liters/sec.	Gallons/min.	0.063 08
Lbs.	Kilograms	2.205
Lbs./foot	Kgs./meter	0.672 0
Lbs./cubic foot	Pounds/cubic inch	1 728

To obtain	*Multiply*	*By*
Lbs./cubic inch	Pounds/cubic foot	5.787×10^{-4}
Lbs./sq. ft.	Feet of water	62.43
Lbs./sq. ft.	Inches of mercury	70.73
Lbs./sq. foot	Inches of water	5.202
Lbs./sq. foot	Kgs./sq. meter	0.204 8
Lbs./sq. inch	Feet of water	0.433 5
Lbs./sq. inch	Inches of mercury	0.491 2
Lbs./sq. inch	Inches of water	0.036 13
Lbs./sq. inch	Kgs./sq. meter	1.422×10^{-3}
Lbs./sq. inch	Ounces/sq. inch	0.062 5
Lbs./million gal.	Grains/U.S. gal.	142.86
Lbs./million gal.	Parts/million	8.345
Meters	Hectometers	100
Meters	Kilometers	10^3
Meters	Microns	10^{-6}
Meters/min.	Feet/min.	0.304 8
Meters/min.	Feet/sec.	18.29
Meters/min.	Kilometers/hr.	16.67
Meters/min.	Miles/hr.	26.82
Meters/sec./sec.	Feet/sec./sec.	0.304 8
Meters/sec./sec.	Kms./hr./sec.	0.277 8
Miles	Kilometers	0.621 4
Miles/hr.	Feet/min.	0.011 36
Miles/hr.	Feet/sec.	0.681 8
Miles/hr.	Kilometers/hr.	0.621 4
Miles/hr.	Miles/min.	60
Miles/hr.	Meters/min.	0.037 28
Miles/hr.	Meters/sec.	2.237
Miles/min.	Feet/sec.	0.011 36
Miles/min.	Meters/sec.	0.037 28
Milligrams	Grams	10^3
Millimeters	Meters	10^3
Minutes	Quadrants (angle)	5 400
Minutes	Radians	3 438
Ounces	Grams	0.035 27
Ounces	Pounds	16
Ounces (avoir.)	Ounces (troy)	1.097 14
Ounces (avoir.)	Pounds (troy)	13.165 7
Ounces (troy)	Grains (troy)	$2.083\ 3 \times 10^{-3}$
Ounces (troy)	Grams	0.032 15

To obtain	Multiply	By
Ounces (troy)	Ounces	0.911 5
Ounces (troy)	Pennyweights (troy)	0.05
Ounces (troy)	Pounds	14.583 3
Ounces (troy)	Pounds (troy)	12
Ounces/sq. inch	Inches of water	0.578 1
Overflow rate	Gallons/min.	8.020 8
Parts/million	Grains/Imp. gal.	14.254
Parts/million	Grains/U.S. gal.	17.118
Parts/million	Grams/liter	1 000
Parts/million	Milligrams/liter	1
Pennyweights (troy)	Grains (troy)	0.041 67
Pennyweights (troy)	Ounces (troy)	20
Pennyweights (troy)	Pounds (troy)	240
Pints (liq.)	Gallons	8
Pints (liq.)	Liters	2.113
Pounds	Grams	2.205×10^{-3}
Pounds	Ounces	0.062 5
Pounds	Quintal, Argentine	101.28
Pounds	Quintal, Brazil	129.54
Pounds	Quintal, Castile, Peru	101.43
Pounds	Quintal, Chile	101.41
Pounds	Quintal, Metric	220.46
Pounds	Quintal, Mexico	101.47
Pounds (avoir.)	Pounds (troy)	0.822 857
Pounds (troy)	Pennyweights (troy)	$4.166 7 \times 10^{-3}$
Pounds (troy)	Pounds	1.215 28
Pounds (troy)	Ounces (troy)	0.083 33
Pounds/cubic foot	Grams/cu. cm.	62.43
Pounds/cubic foot	Grams/liter	0.062 427
Pounds/cubic inch	Grams/cu. cm.	0.036 13
Pounds/sq. inch	Pounds/sq. foot	6.945×10^{-3}
Pounds/inch	Grams/cm.	5.600×10^{-3}
Pounds of water	Gallons-water	8.345 3

To obtain	Multiply	By
Pounds/1 000 gals.	Grams/liter	8.345
Quadrants	Radians	0.637
Quadrants	Revolutions	4
Quarts (liq.)	Liters	1.057
Quarts (liq.)	Gallons	4
Radians	Minutes (angle)	2.909×10^{-4}
Radians	Quadrants (angle)	1.571
Radians	Revolutions	6.283
Radians	Seconds (angle)	4.848×10^{-6}
Radians/sec.	Revolutions/min.	0.104 7
Radians/sec.	Revolutions/sec.	6.283
Rads./sec./sec.	Revolutions/ min./min.	1.745×10^{-8}
Radians/sec./sec.	Revolutions/ sec./sec.	6.283
Revolutions/min.	Radians/sec.	9.549
Revolutions/min.	Revolutions/sec.	60
Revs./min./min.	Radians/sec./sec.	573.0
Revs./min./min.	Revolutions/ sec./sec.	3 600
Revolutions/sec.	Radians/sec.	0.159 2
Revolutions/sec.	Revolutions/min.	0.016 67
Revs./sec./sec.	Radians/sec./sec.	0.159 2
Revs./sec./sec.	Revolutions/ min./min.	2.778×10^{-4}
Sheets	Quires	25
Sheets	Reams	500
Square feet	Square centimeters	1.076×10^{-3}
Square inches	Square centimeters	0.155 0
Sq. ft./gal./min.	$\dfrac{1}{\text{Overflow rate (ft./hr.)}}$	8.020 8
Square feet	Hectares	10^5
Tons (metric)	Ounces	2.835×10^{-5}
Tons (metric)	Pounds (troy)	$3.732\ 4 \times 10^{-4}$
Tons (long)	Ounces	2.790×10^{-5}
Tons (long)	Pounds (troy)	$3.673\ 5 \times 10^{-4}$
Tons (short)	Kilograms	1.102×10^{-3}
Tons (short)	Pounds	0.000 5

To obtain	Multiply	By
Tons (short)	Pounds (troy)	$4.114\,3 \times 10^{-4}$
Tons water/24 hrs.	Gallons water/min.	6.008 6
U.S. Gallons	Gallons-Imperial	1.200 95
Watts	Hectowatts	100
Watts	Horsepower	745.7
Watts	Kilowatts	10^3
Yards	Feet	1/3
Yards	Kilometers	1 094
Yards	Meters	1.094
Yards	Miles	1 760

APPENDIX

APPENDIX

Introduction and Explanation of the Metric Conversion Act Approved on December 23, 1975

The Metric Conversion Act of 1975 was considered and passed by the House on September 5, 1975. On December 8, 1975, it was considered and passed by the Senate and amended in lieu of metric bill S100. The House concurred with the Senate amendment on December 11, 1975. The Metric Conversion Act was signed by the President on December 23, 1975.

Three metric bills were considered by Congress in 1975: S.100, S.1882, and H.R. 8674. H.R. 8674 was approved with amendments.

The purpose of the bill is to establish a national metric policy and to outline the method for developing and organizing the efforts necessary for metric conversion in each area of our economy. The bill contains a brief history of America's metric legacy and explains its previous commitments to accept the metric system of measurement, i.e. America's 1875 signing of the Treaty of the Meter and the Metric Act of July 28, 1866.

In essence, the bill declares that the policy of the United States shall be to plan for the coordinate efforts toward the increased use of the metric system. The vehicle which shall move the nation toward metric conversion is the United States Metric Board as established by the Metric Act of 1975. The bill outlines the procedure for establishing the Metric Board and defines its mission with substantial detail and direction. Although Sec. 3 of the bill contains the phrase "voluntary conversion to the metric system," the specific provision in the bill, especially those concerning the mission of the Metric Board, leave no doubt as to our nation's commitment to adopt the metric system of measurement.

The complete text of the Metric Act of 1975 is followed by the author's comments and a copy of the Public Law 90-472, which made the studies leading to the bill possible.

Public Law 94-168
94th Congress, H. R. 8674
December 23, 1975

An Act

To declare a national policy of coordinating the increasing use of the metric system in the United States, and to establish a United States Metric Board to coordinate the voluntary conversion to the metric system.

Be it enacted by the Senate and House of Representatives of the United States of America in Congress assembled, That this Act may be cited as the "Metric Conversion Act of 1975".

Metric Conversion Act of 1975. 15 USC 205a note. 15 USC 205a.

SEC. 2. The Congress finds as follows:

(1) The United States was an original signatory party to the 1875 Treaty of the Meter (20 Stat. 709), which established the General Conference of Weights and Measures, the International Committee of Weights and Measures and the International Bureau of Weights and Measures.

(2) Although the use of metric measurement standards in the United States has been authorized by law since 1866 (Act of July 28, 1866; 14 Stat. 339), this Nation today is the only industrially developed nation which has not established a national policy of committing itself and taking steps to facilitate conversion to the metric system.

SEC. 3. It is therefore declared that the policy of the United States shall be to coordinate and plan the increasing use of the metric system in the United States and to establish a United States Metric Board to coordinate the voluntary conversion to the metric system.

15 USC 205b.

SEC. 4. As used in this Act, the term—

(1) "Board" means the United States Metric Board, established under section 5 of this Act;

(2) "engineering standard" means a stan-

Definitions. 15 USC 205c.

dard which prescribes (A) a concise set of conditions and requirements that must be satisfied by a material, product, process, procedure, convention, or test method; and (B) the physical, functional, performance and/or conformance characteristics thereof;

(3) "international standard or recommendation" means an engineering standard or recommendation which is (A) formulated and promulgated by an international organization and (B) recommended for adoption by individual nations as a national standard; and

(4) "metric system of measurement" means the International System of Units as established by the General Conference of Weights and Measures in 1960 and as interpreted or modified for the United States by the Secretary of Commerce.

United States Metric Board. Establishment. 15 USC 205d. Membership.

SEC. 5. (a) There is established, in accordance with this section, an independent instrumentality to be known as a United States Metric Board.

(b) The Board shall consist of 17 individuals, as follows:

(1) the Chairman, a qualified individual who shall be appointed by the President, by and with the advice and consent of the Senate;

(2) sixteen members who shall be appointed by the President, by and with the advice and consent of the Senate, on the following basis—

(A) one to be selected from lists of qualified individuals recommended by engineers and organizations representative of engineering interests;

(B) one to be selected from lists of qualified individuals recommended by scientists, the scientific and technical community, and organizations representative of scientists and technicians;

(C) one to be selected from a list of qualified individuals recommended by the National Association of Manufacturers or its successor;

(D) one to be selected from lists of qualified individuals recommended by the United States Chamber of Commerce, or its successor, retailers, and other commercial organizations;

(E) two to be selected from lists of qualified individuals recommended by the American Federation of Labor and Congress of Industrial organizations or its successor, who are representative of workers directly affected by metric conversion, and by other organizations representing labor;

(F) one to be selected from a list of qualified individuals recommended by the National Governors Conference, the National Council of State Legislatures, and organizations representative of State and local government;

(G) two to be selected from lists of qualified individual recommended by organizations representative of small business;

(H) one to be selected from lists of qualified individuals representative of the construction industry;

(I) one to be selected from a list of qualified individuals recommended by the National Conference on Weights and Measures and standards making organizations;

(J) one to be selected from lists of qualified individuals recommended by educators, the educational community, and organizations representative of educational interests; and

(K) four at-large members to represent consumers and other interests deemed suitable by the President and who shall be qualified individuals.

As used in this subsection, each "list" shall include the names of at least three individuals for each applicable vacancy. The terms of office of the members of the Board first taking office shall expire as designated by the President at the time of nomination; five at the end of the 2d year; five at the end of the

Term of office.

4th year; and six at the end of the 6th year. The term of office of the Chairman of such Board shall be 6 years. Members, including the Chairman, may be appointed to an additional term of 6 years, in the same manner as the original appointment. Successors to members of such Board shall be appointed in the same manner as the original members and shall have terms of office expiring 6 years from the date of expiration of the terms for which their predecessors were appointed. Any individual appointed to fill a vacancy occurring prior to the expiration of any term of office shall be appointed for the remainder of that

Quorum. term. Beginning 45 days after the date of incorporation of the Board, six members of such Board shall constitute a quorum for the transaction of any function of the Board.

(c) Unless otherwise provided by the Congress, the Board shall have no compulsory powers.

(d) The Board shall cease to exist when the Congress, by law, determines that its mission has been accomplished.

Policy implementation. 15 USC 205e.
SEC. 6. It shall be the function of the Board to devise and carry out a broad program of planning, coordination, and public education, consistent with other national policy and interests, with the aim of implementing the policy set forth in this Act. In carrying out this program, the Board shall—

(1) consult with and take into account the interests, views, and conversion costs of United States commerce and industry, including small business; science; engineering; labor; education; consumers; government agencies at the Federal, State, and local level; nationally recognized standards developing and coordinating organizations; metric conversion planning and coordinating groups; and such other individuals or groups as are considered appropriate by the Board to the carrying out of the purposes of this Act. The Board shall take into account activities underway in the private and public sectors, so as not to duplicate unnecessarily such activities;

(2) provide for appropriate procedures whereby various groups, under the auspices of the Board, may formulate, and recommend or suggest, to the Board specific programs for coordinating conversion in each industry and segment thereof and specific dimensions and configurations in the metric system and in other measurements for general use. Such programs, dimensions, and configurations shall be consistent with (A) the needs, interests, and capabilities of manufacturers (large and small), suppliers, labor, consumers, educators, and other interested groups, and (B) the national interest;

(3) publicize, in an appropriate manner, proposed programs and provide an opportunity for interested groups or individuals to submit comments on such programs. At the request of interested parties, the Board, in its discretion, may hold hearings with regard to such programs. Such comments and hearings may be considered by the Board; *Comments and hearings.*

(4) encourage activities of standardization organizations to develop or revise, as rapidly as practicable, engineering standards on a metric measurement basis, and to take advantage of opportunities to promote (A) rationalization or simplification of relationships, (B) improvements of design, (C) reduction of size variations, (D) increases in economy, and (E) where feasible, the efficient use of energy and the conservation of natural resources;

(5) encourage the retention, in new metric language standards, of those United States engineering designs, practices, and conventions that are internationally accepted or that embody superior technology;

(6) consult and cooperate with foreign governments, and intergovernmental organizations, in collaboration with the Department of State, and, through appropriate member bodies, with private international organizations, which are or become concerned with the encouragement and *Consultation and cooperation.*

coordination of increased use of metric measurement units or engineering standards based on such units, or both. Such consultation shall include efforts, where appropriate, to gain international recognition for metric standards proposed by the United States, and, during the United States conversion, to encourage retention of equivalent customary units, usually by way of dual dimensions, in international standards or recommendations;

Public information and education programs.

(7) assist the public through information and education programs, to become familiar with the meaning and applicability of metric terms and measures in daily life. Such programs shall include—

(A) public information programs conducted by the Board, through the use of newspapers, magazines, radio, television, and other media, and through talks before appropriate citizens' groups, and trade and public organizations;

(B) counseling and consultation by the Secretary of Health, Education, and Welfare; the Secretary of Labor; the Administrator of the Small Business Administration; and the Director of the National Science Foundation, with educational associations, State and local educational agencies, labor education committees, apprentice training committees, and other interested groups, in order to assure (i) that the metric system of measurement is included in the curriculum of the Nation's educational institutions, and (ii) that teachers and other appropriate personnel are properly trained to teach the metric system of measurement;

(C) consultation by the Secretary of Commerce with the National Conference of Weights and Measures in order to assure that State and local weights and measures officials are (i) appropriately involved in metric conversion activities and (ii) assisted in their ef-

forts to bring about timely amendments to
weights and measures laws; and

(D) such other public information activities, by any Federal agency in support of this
Act, as relate to the mission of such agency;

(8) collect, analyze, and publish information
about the extent of usage of metric measurements; evaluate the costs and benefits of metric
usage; and make efforts to minimize any adverse effects resulting from increasing metric
usage;

(9) conduct research, including appropriate
surveys; publish the results of such research;
and recommend to the Congress and to the
President such action as may be appropriate to
deal with any unresolved problems, issues, and
questions associated with metric conversion, or
usage, such problems, issues, and questions may
include, but are not limited to, the impact on
workers (such as costs of tools and training)
and on different occupations and industries,
possible increased costs to consumers, the impact on society and the economy, effects on
small business, the impact on the international
trade position of the United States, the appropriateness of and methods for using procurement by the Federal Government as a means to
effect conversion to the metric system, the proper conversion or transition period in particular
sectors of society, and consequences for national defense;

Surveys. Recommendations to Congress and President.

(10) submit annually to the Congress and to
the President a report on its activities. Each
such report shall include a status report on the
conversion process as well as projections for the
conversion process. Such report may include
recommendations covering any legislation or executive action needed to implement the programs of conversion accepted by the Board.
The Board may also submit such other reports
and recommendations as it deems necessary;
and

Report to Congress and President.

Report
to Con-
gress and
Pres-
ident.

(11) submit to the Congress and to the President, not later than 1 year after the date of enactment of the Act making appropriations for carrying out this Act, a report on the need to provide an effective structural mechanism for converting customary units to metric units in statutes, regulations, and other laws at all levels of government, on a coordinated and timely basis, in response to voluntary conversion programs adopted and implemented by various sectors of society under the auspices and with the approval of the Board. If the Board determines that such a need exists, such report shall include recommendations as to appropriate and effective means for establishing and implementing such a mechanism.

SEC. 7. In carrying out its duties under this Act, the Board may—

Commit-
tees, es-
tablish-
ment.
15 USC
205f.

(1) establish an Executive Committee, and such other committees as it deems desirable;

(2) establish such committees and advisory panels as it deems necessary to work with the various sectors of the Nation's economy and with Federal and State governmental agencies in the development and implementation of detailed conversion plans for those sectors. The Board may reimburse, to the extent authorized by law, the members of such committees;

Hearings.

(3) conduct hearings at such times and places as it deems appropriate;

Con-
tracts.

(4) enter into contracts, in accordance with the Federal Property and Administrative Services Act of 1949, as amended (40 U.S.C. 471 et seq.), with Federal or State agencies, private firms, institutions, and individuals for the conduct of research or surveys, the preparation of reports, and other activities necessary to the discharge of its duties;

(5) delegate to the Executive Director such authority as it deems advisable; and

(6) perform such other acts as may be

necessary to carry out the duties prescribed by this Act.

SEC. 8. (a) The Board may accept, hold, administer, and utilize gifts, donations, and bequests of property, both real and personal, and personal services, for the purpose of aiding or facilitating the work of the Board. Gifts and bequests of money, and the proceeds from the sale of any other property received as gifts or bequests, shall be deposited in the Treasury in a separate fund and shall be disbursed upon order of the Board. *Gifts and bequests. 15 USC 205g.*

(b) For purpose of Federal income, estate, and gift taxation, property accepted under subsection (a) of this section shall be considered as a gift or bequest to or for the use of the United States.

(c) Upon the request of the Board, the Secretary of the Treasury may invest and reinvest, in securities of the United States, any moneys contained in the fund authorized in subsection (a) of this section. Income accruing from such securities, and from any other property accepted to the credit of such fund, shall be disbursed upon the order of the Board.

(d) Funds not expended by the Board as of the date when it ceases to exist, in accordance with section 5(d) of this Act, shall revert to the Treasury of the United States as of such date. *Unexpended funds.*

SEC. 9. Members of the Board who are not in the regular full-time employ of the United States shall, while attending meetings or conferences of the Board or while otherwise engaged in the business of the Board, be entitled to receive compensation at a rate not to exceed the daily rate currently being paid grade 18 of the General Schedule (under section 5332 of title 5, United States Code), including traveltime. While so serving, on the business of the Board away from their homes or regular places of business, members of the Board may be allowed travel expenses, including per diem in lieu of subsistence, as authorized by section 5703 of title 5, United States Code, for persons employed intermittently in the Government service. Payments under this section shall not render members of the Board *Compensation. 15 USC 205h.* *5 USC 5332 note. Travel expenses.*

employees or officials of the United States for any purpose. Members of the Board who are in the employ of the United States shall be entitled to travel expenses when traveling on the business of the Board.

Execu-
tive Di-
rector,
appoint-
ment.
15 USC
205i.

SEC. 10. (a) The Board shall appoint a qualified individual to serve as the Executive Director of the Board at the pleasure of the Board. The Executive Director, subject to the direction of the Board, shall be responsible to the Board and shall carry out the metric conversion program, pursuant to the provisions of this Act and the policies established by the Board.

5 USC
5101
et seq.
5 USC
5331.

(b) The Executive Director of the Board shall serve full time and be subject to the provisions of chapter 51 and subchapter III of chapter 53 of title 5, United States Code. The annual salary of the Executive Director shall not exceed level III of the Executive Schedule under section 5314 of such title.

(c) The Board may appoint and fix the compensation of such staff personnel as may be necessary to carry out the provisions of this Act in accordance with the provisions of chapter 51 and subchapter III of chapter 53 of title 5, United States Code.

Experts
and con-
sultants.

(d) The Board may (1) employ experts and consultants or organizations thereof, as authorized by section 3109 of title 5, United States Code; (2) compensate individuals so employed at rates not in excess of the rate currently being paid grade 18 of the General Schedule under section 5332 of such title, including traveltime; and (3) may allow such individuals, while away from their homes or regular places of business, travel expenses (including per diem in lieu of subsistence) as authorized by section 5703 of such title 5 for persons in the Government service employed intermittently: *Provided, however,* That contracts for such temporary employment may be renewed annually.

SEC. 11. Financial and administrative services, including those related to budgeting, accounting, financial reporting, personnel, and procurement, and such other staff services as may be needed by the Board, may be obtained by the Board from the Secretary of Commerce or other appropriate sources in the Federal Government. Payment for such services shall be made by the Board, in advance or by reimbursement, from funds of the Board in such amounts as may be agreed upon by the Chairman of the Board and by the source of the services being rendered. *Financial and administrative services. 15 USC 205j.*

SEC. 12. There are authorized to be appropriated such sums as may be necessary to carry out the provisions of this Act. Appropriations to carry out the provisions of this Act may remain available for obligation and expenditure for such period or periods as may be specified in the Acts making such appropriations. *Appropriation authorization. 15 USC 205k.*

Approved December 23, 1975.

Major Provisions of the Bill, with Comments

The Metric Conversion Act of 1975 (The Bill) is founded on three years of studies, surveys, and analyses by the National Bureau of Weights and Measures in accordance with the Metric Study Act of 1968. The basic recommendation resulting from the 1968 study is a systematic, nationally coordinated U.S. changeover to the metric system of measurement to take place over a period of ten years. Other recommendations based on the study and forwarded to Congress are:

1. That the United States change to the International Metric System deliberately and carefully;
2. That this be done through a coordinated national program;
3. That the Congress assign the responsibility for guiding the change, and anticipating the kinds of special problems described in the report, to a central coordinating body responsive to all sectors of our society;
4. That within this guiding framework, detailed plans and timetables be worked out by these sectors themselves;
5. That early priority be given to educating every American schoolchild and the public at large to think in metric terms;
6. That immediate steps be taken by the Congress to foster U.S. participation in international standards activities;
7. That the Congress, after deciding on a plan for the nation, establish a target date ten years ahead, by which time the U.S. will have become predominantly, though not exclusively, metric;
8. That there be a firm government commitment to this goal.

Comments

It is obvious that through clearly stated intentions, our government hopes to lead us to a metric world as painlessly and efficiently as possible. Flexibility, time, and technical assistance are offered by the bill. The bill seems to focus almost exclusively on the formation of the United States Metric Board. Although the United States Metric Board has no compulsory powers in itself, its responsibility to Congress as the major vehicle for effecting an equitable metric changeover is clearly stated.

The United States Metric Board shall consist of seventeen members representing every sector of our economy, i.e. private and public, labor and management, scientific and technical, education and engineering. The Board is charged with a broad program of planning, coordination, and public education. It shall also engage in close consultation and planning of a detailed nature with each sector of our economy such as small business, science, engineering, labor, education, manufacturing and consumerism. The Board is responsible for encouraging technical organizations to metricate engineering standards as rapidly as is practical while encouraging the retention of metrically redefined American engineering designs, practices, and conventions that are now internationally accepted or that involve superior technology. The Board shall arrange for cooperation and consultation with foreign governments and international organizations concerned with metric measurements. Such activity shall include efforts to influence international measurement policy.

Educators are specifically affected by the bill, i.e. (1) they are to assure that the metric system of measurement is included in the curriculum of the nation's educational institutions and (2) that teachers and other appropriate personnel are properly trained to teach the metric system. The Board shall assist the public through information and education programs via the mass media.

The Board may employ experts and consultants. It may conduct research and surveys and publish the results. It may recommend to Congress and the President any action or legislation needed (1) to alleviate unresolved problems, issues, and questions concerning metric changeover or (2) to implement the programs of conversion accepted by the Board.

In conclusion, it is the Board's responsibility to develop and implement detailed metric conversion plans for all sectors of the economy including Federal and State governmental agencies. The bill is a strong one. Its strength lies in the provisions which outline and define the development of the United States Metric Board and its mission.

Public Law 90-472

Public Law 90-472, which made the studies leading to the bill possible, is brief and explicit in its intent. It is presented here intact.

An Act

To authorize the Secretary of Commerce to make a study to determine advantages and disadvantages of increased use of the metric system in the United States.

Be it enacted by the Senate and House of Representatives of the United States of America in Congress assembled, That the Secretary of Commerce is hereby authorized to conduct a program of investigation, research, and survey to determine the impact of increasing worldwide use of the metric system on the United States; to appraise the desirability and practicability of increasing the use of metric weights and measures in the United States; to study the feasibility of retaining and promoting by international use of dimensional and other engineering standards based on the customary measurement units of the United States; and to evaluate the costs and benefits of alternative courses of action which may be feasible for the United States.

SECTION 2. In carrying out the program described in the first section of this Act, the Secretary, among other things, shall—

(1) investigate and appraise the advantages and disadvantages to the United States in international trade and commerce, and in military and other areas of international relations, of the increased use of an internationally standardized system of weights and measures;

(2) appraise economic and military advantages and disadvantages of the increased use of the metric system in

the United States or of the increased use of such system in specific fields and the impact of such increased use upon those affected;

(3) conduct extensive comparative studies of the systems of weights and measures used in educational, engineering, manufacturing, commercial, public, and scientific areas, and the relative advantages and disadvantages, and degree of standardization of each in its respective field;

(4) investigate and appraise the possible practical difficulties which might be encountered in accomplishing the increased use of the metric system of weights and measures generally or in specific fields or areas in the United States;

(5) permit appropriate participation by representatives of United States industry, science, engineering, and labor, and their associations, in the planning and conduct of the program authorized by the first section of this Act, and in the evaluation of the information secured under such program; and

(6) consult and cooperate with other government agencies, Federal, State, and local, and, to the extent practicable, with foreign governments and international organizations.

SECTION 3. In conducting the studies and developing the recommendations required in this Act, the Secretary shall give full consideration to the advantages, disadvantages, and problems associated with possible changes in either the system of measurement units or the related dimensional and engineering standards currently used in the United States, and specifically shall—

(1) investigate the extent to which substantial changes in the size, shape, and design of important industrial products would be necessary to realize the benefits which might result from general use of metric units of measurement in the United States;

(2) investigate the extent to which uniform and accepted engineering standards based on the metric system of measurement units are in use in each of the fields under study and compare the extent of such use and the utility and degree of sophistication of such metric standards with those in use in the United States; and

(3) recommend specific means of meeting the practical

difficulties and costs in those areas of the economy where any recommended change in the system of measurement units and related dimensional and engineering standards would raise significant practical difficulties or entail significant costs of conversion.

SECTION 4. The Secretary shall submit to the Congress such interim reports as he deems desirable, and within three years after the date of the enactment of this Act, a full and complete report of the findings made under the program authorized by this Act, together with such recommendations as he considers to be appropriate and in the best interests of the United States.

SECTION 5. From funds previously appropriated to the Department of Commerce, the Secretary is authorized to utilize such appropriated sums as are necessary, but not to exceed $500,000, to carry out the purposes of this Act for the first year of the program.

SECTION 6. This Act shall expire thirty days after the submission of the final report pursuant to section 3. Approved Aug. 9, 1968.

BIBLIOGRAPHY
AND LIST
OF REFERENCES

BIBLIOGRAPHY
AND LIST OF REFERENCES

"All You Will Need to Know About Metric," a handout sheet which details all the average person will need to learn. By National Bureau of Standards.

"Are Inches, Pints, Pounds on the Way Out in the U.S.?" *U.S. News*, 71:73-74, October 11, 1971.

Asimov, I. "How Many Inches in a Mile?" With editorial comment, *Saturday Evening Post*, 243:96-98, Winter, 1971.

"Assignment Cards on Metric Units of Measurement," Junior Set, Middle Set and Senior Set. Department of Education, Wellington, New Zealand.

Bailey, J.H. "Farewell to the Barleycorn Inch," *Science Digest*, 69:26-30, January, 1971.

Berger, Melvin. *For Good Measure: The Story of Modern Measurement*. New York: McGraw-Hill Book Company, 1969.

Bibliography on the Metric System: Instructional Materials. Indiana Department of Public Instruction, 1974. 35 pp.

Bibliography on the Metric System: Periodical Articles. Indiana Department of Public Instruction, 1974. 9 pp.

Brief History of Measurement System (NBS Special Publication 304A, revised October, 1972).

Canadian Standards Association. "The International System of Units (SI): An Outline of Canadian Usage." Special paper, June, 1973.

"Cold Now? Wait 'Til We Go Metric!" *Buffalo Evening News*, February 1, 1975, p. c-6.

Conversion Materials, Metric Conversion Board, Australia, 1973.

Copeland, Richard W. *Mathematics and The Elementary Teacher*. W. B. Saunders Company, 1972.

Copy of H. R. 5749, which was prepared by the Department of Commerce. It is similar to the legislation that was passed in 1972 by the U.S. Senate and may well be similar to legislation that may be enacted by the 93rd Congress.

Department of Education. *Metric Units of Measurements*. Curriculum Development Unit Bulletin, No. 50. New Zealand, 1972.

DeSimone, D. V. "Moving to Metric Makes Dollars and Sense with Tables," *Harvard Business Review*, 50:100-111, January, 1972.

"A Detailed List of Conversion Factors for Metric and Customary Units" (reprinted from NBS Miscellaneous Publication 286).

Donovan, Frank. *Prepare Now for a Metric Future*. New York: Weybright and Talley, 1970.

Dumesnils, Maurice Danloux. *The Metric System: A Critical Study of Its Principles and Practice*. London: The Athlone Press, 1969.

Edson, L. "New Dimensions for Practically Everything: Metrication," *American Education*, 8:10-14, April, 1972.

"English-Metric Conversion Calculator," Union Carbide Corporation, Tuxedo, New York.

"First Step Toward U.S. Adoption of the Metric System," *U.S. News*, 69:88, October 19, 1970.

Ford Motor Company Metric Materials.

Government Printer, Visual Production Unit, Department of Education; Poster/Chart No. 72/801. New Zealand.

Grohame, J.A. Maxtone. "Alas, Poor England! Metrication is Coming." *Sports Illustrated*, 33:47, August 24, 1970.

Hawkins, Vincent J. "Teaching the Metric System as Part of Compulsory Conversion in the United States," *The Arithmetic Teacher*, May, 1973.

Holden, C. "NBS Urges 10-year Metric Conversion Plan," *Science*, 173:613, August 13, 1971.

"Idea Whose Time Has Come," *Science News*, 100:92, August 7, 1971.

Ingalls, Walter Renton. *Modern Weights and Measures*. New York: published by American Institute of Weights and Measures, 1937.

———. *Systems of Weights and Measures*. New York: published by American Institute of Weights and Measures, 1945.

Izzi, John. "A Right Now Project: How to Get Ready to Go Metric in Your School District," *American School Board Journal*, July, 1973.

Johnson, Julia. *Metric System.* New York: H. W. Wilson Company, 1926.

Jones, Phillip G. "Metrics: Your schools will be teaching it and you'll be living it—very, very, very soon," "Metrics Will Be Published on Schools by Dollar Considerations, but the system makes sense," "(Honest; it's easy) Why We Adults can't (and shouldn't) Escape A Metric World," "What it will cost to go Metric," *American School Board Journal*, July, 1973.

Kendig, F. "Coming of the Metric System," *Saturday Review*, 55.40-44, November 25, 1972.

Kennelly, A. E. *Vestiges of Pre-Metric Weights and Measures.* New York: The Macmillan Company, 1928.

Kerr, K. F. *et al. Metric Weights and Measures for Schools in Programmed Form. Part I, Introduction Length and Area, Part II, Volume, Capacity and Weight.* W. Foulshaw and Company, Ltd., London, New York, Toronto, Sidney, Capetown, 1970. 33 pages each.

"Lets Go Metric," *Life,* 71:42, October 18, 1971.

Lodgdon, G. "Coming: The Modernized Metric System," *Farm Journal,* 95:57-58, February, 1971.

Manchester, H. "Here Comes the Meter," *Readers Digest*, 100:19-20, April, 1972.

Mattoon, R. W. "Metric System: Status of Adoption by the United States," AAAS Symposium, *Science,* December 28, 1970.

Metric Materials Catalog, Dick Blick Company, P.O. Box 1267, Galesburg, Illinois, 61401.

"Meter Readers: Teaching the Metric System," *Newsweek,* 81:65, May 7, 1973.

Metric Board Pamphlets, London, England. "Going Metric," "The Kilogram," "The Liter," "Metric Weights and Measures," "The Meter," "Temperature."

"Metric Measures: Closer Than you Think; with Tables," *Successful Farmer,* 70:38, September, 1972.

Metric Memo, No. 1, *Everyday Units,* Government Printer, New Zealand, 1972.

"Metric System," *McGraw-Hill Encyclopedia of Science and Technology*, Vol. 8, pp. 404-405, 1971.

Metric Practice Guide, American Society for Testing and Materials, E 380–, 1972.

"Metrication in America," *Science American*, 223:520, 1970.

"Metrication's Justifications," *Consumer Bulletin*, 55:34-36, January, 1972.

"Metrikit Mini-Course," Villa Park, Illinois: Larry Harkness Company, May, 1973.

Meyer, C. "Metrication of the American Housewife," *McCalls*, 99:53, March, 1972.

"Mounting Pressure to go Metric," *Business Week*, July 24, 1971, pp. 54-56.

"News From the World of Space Exploration: Metric System for NASA," *Space World*, 85:47, January, 1971.

Offenbacker, E. L. "Metric Time History," *Physics Today*, 26:114, January, 1973.

"On the Metric Road," *Science American*, 226:48-49, May, 1972.

Page, C.H. "Ambiguities in the Use of Unit Names," *Science*, 179:873-875, March 2, 1973.

A Pocket Metric Conversion Card (NBS Special Publication 365), giving approximate conversion factors for the most commonly used units.

"Preliminary Report: NBS Study of Changeover Problems," *Science News*, 96:297-298, October 4, 1969.

Ritchie-Colder, Lord. "Conservation to the Metric System," *Science American*, 223:17-25, July, 1970.

Schwartz, Allen A. and Ann Edson. "Think Metric," a kit of four sound/color filmstrips for grades 4-6. Educational Activities, Inc. Freeport, New York.

Shapley, D. "Metric A Go-Go," *Science*, 175:1346, March 24, 1972.

Short, Pat. *Learn Metric, Think Metric, Go Metric!* East York Board of Education, Toronto, Ontario, 1974. 20 pp.

"Should the U.S. Go Metric?" *Senior Scholastic*, 96:111-112, March 16, 1970.

"Soon It May Be Give a Centimeter and Take a Kilometer," *Nation's Business*, 59:88-89, April, 1971.

Stans, H. "Got to Get Organized: Proposals of Maurice H. Stans," *Newsweek*, 78:23, August 9, 1971.

A Summary of the Final Report of the U.S. Metric Study (reprint from NBS Technical News Bulletin, September, 1971).

"Take Me to Your Liter," *Nation's Business,* 58:18, June, 1970.

"Ten Years to Metric: Australian Conversion," *Science News,* 97:340, April 4, 1970.

"This Month's Feature: The Metric System Controversy," *Congressional Digest,* 50:289-314, December, 1971. April, 1972.

The Arithmetic Teacher. NCTM. Entire Issue, April, 1973.

"The Metric System," Chart Set. The Instructor Publications, Inc., Danville, New York, 1971.

"Toward a Metric System," National Metric Education Center, *Intellect,* 101:285, February, 1973.

"Toward a Metric U.S.," *Science American,* 225:76, September, 1971.

"Toward a 1,024-joule Hamburger," *Science American,* 226:56.

Trueblood, Cecil R. *Metric Measurement Activities and Bulletin Boards.* Danville, New York: Instructor Publications, 1973.

U.S. Department of Commerce. *A Metric America: A Decision Whose Time Has Come.* Washington, D.C.; Government Printing Office, 1971.

U.S. Department of Commerce, National Bureau of Standards.

Units of Weights and Measures. Miscellaneous Publications No. 286. Issued May 1967.

————. *Special Publication 304A,* Rev. 1972.

————. *Special Publication 330,* 1972 Edition.

————. "NBS Policy for Usage of SI Units," *Technical News Bulletin,* 57 No. 6 (June 1973).

"U.S. Inches Closer to the Meter," *Chicago Tribune,* June 6, 1971.

"Using Metric System: Drastic Change for Cooks," *Buffalo Evening News,* March 22, 1973.

Wesner, Gordon E. and M. Potter. "History and Explanation of the Metric System," a 13-page paper written for Beloit Tool.

INDEX

INDEX